Sociocultural Explorations of Science Education

Volume 32

The series is unique in focusing on the publication of scholarly works that employ social and cultural perspectives as foundations for research and other scholarly activities in the three fields implied in its title: science education, education, and social studies of science.

The aim of the series is to promote transdisciplinary approaches to scholarship in science education that address important topics in the science education including the teaching and learning of science, social studies of science, public understanding of science, science/technology and human values, science and literacy, ecojustice and science, indigenous studies and science and the role of materiality in science and science education. Cultural Studies of Science Education, the book series explicitly aims at establishing such bridges and at building new communities at the interface of currently distinct discourses. In this way, the current almost exclusive focus on science education on school learning would be expanded becoming instead a focus on science education as a cultural, cross-age, cross-class, and cross-disciplinary phenomenon.

The book series is conceived as a parallel to the journal Cultural Studies of Science Education, opening up avenues for publishing works that do not fit into the limited amount of space and topics that can be covered within the same text. Book proposals for this series may be submitted to the Publishing Editor: Claudia Acuna E-mail: Claudia.Acuna@springer.com

Alberto Bellocchi · Julia Hill · Louisa Tomas ·
Tanya Doyle

Emotion Research in Science Teacher Education

A Systematic Review

Springer

Alberto Bellocchi
Queensland University of Technology
Kelvin Grove, QLD, Australia

Louisa Tomas
James Cook University
Douglas, QLD, Australia

Julia Hill
RMIT University
Melbourne, VIC, Australia

University of Melbourne
Parkville, VIC, Australia

Tanya Doyle
James Cook University
Douglas, QLD, Australia

ISSN 2731-0248 ISSN 2731-0256 (electronic)
Sociocultural Explorations of Science Education
ISBN 978-3-032-07710-3 ISBN 978-3-032-07711-0 (eBook)
https://doi.org/10.1007/978-3-032-07711-0

This Springer imprint is published by the registered company Springer Nature Switzerland AG
The registered company address is: Gewerbestrasse 11, 6330 Cham, Switzerland

Preface

Science education has turned its attention to the study of emotions for more than 15 years. One sub-set of this work has addressed the emotions of preservice and inservice science teacher education and professional development. This systematic review outlines the landscape of this body of work. Extending from the earliest time available in research databases to 2025, we discuss major research topics, theories, methodologies, and empirical outcomes of the reviewed studies. The outcomes of the review highlight a diverse set of focal topics for empirical research that science educators have addressed in relation to science teacher emotions. Major theories and methodologies that dominate current research are identified. Drawing on the affordances and limitations of existing knowledge, directions for future research are discussed to advance new directions and support existing trends.

Kelvin Grove, Australia — Alberto Bellocchi
Melbourne, Australia — Julia Hill
Douglas, Australia — Louisa Tomas
Douglas, Australia — Tanya Doyle

Acknowledgement

Initial work began under the auspices of an Australian Research Council Discovery Early Career Grant (DE160101053) awarded to Alberto Bellocchi. Any opinions, findings, and conclusions or recommendations expressed in this article are those of the author and do not necessarily reflect the views of the Australian Research Council.

Contents

Chapter 1
Introduction

"How can teachers' emotions become a legitimate topic in science education as well as in our efforts for science curricular reform? Further, how can educational institutions (e.g., universities and schools) and elementary school science teachers themselves support their personal and professional emotional growth?" (Zembylas, 2001, p. 101). Over twenty years ago, Zembylas (2001) posed the above questions about emotions and science teaching. Although his focus rests on elementary science teaching, subsequent work in high school, early childhood, and university teacher education contexts enables us to answer *yes*, emotions are an important topic of research in science education and they have been associated with much more than reform efforts and professional growth (e.g., Bellocchi, et al., 2018b; Davis & Bellocchi, 2018; Ritchie & Tobin, 2018; Sinatra et al., 2014). Emotions play a pivotal role in teaching and learning, identity development, agency, classroom interactions, professional learning, and wellbeing and retention, but despite growing recognition of the importance of emotions in education, there is a scarcity of publications that focus specifically on emotion in this context. Indeed, the topic of emotion–in any schooling context–is not identified in recently published reviews that have examined research trends in the field of science education (Lin et al., 2025; Odden et al., 2021; Tosun, 2024). This is an interesting omission given that these reviews report that science teacher education and professional development remain among the most researched topics in science education for over 100 years.

Odden et al. (2021) examine how science education has changed over the last 100 years by analysing over 5,500 papers published in *Science Education* from 1922 to 2019 using natural language processing (specifically, Latent Dirichlet Allocation). The authors identify 21 distinct topics grouped into three clusters: science content, teaching-focused, and student-focused. Their analysis reveals interesting shifts in research foci over time. During the 1920s–1950s, science education research is focused primarily on science in everyday life and teacher preparation; the 1970s–1990s reveals the rise of educational psychology, problem-solving, and quantitative studies of motivation and interest; whereas research conducted from the year 2000

A. Bellocchi et al., *Emotion Research in Science Teacher Education*, Sociocultural Explorations of Science Education 32, https://doi.org/10.1007/978-3-032-07711-0_1

to the present emphasises scientific practices, argumentation, identity, and discourse analyses. Teaching-focused topics dominate mid-century research, while student-focused topics surge post-1970. Recent years show strong growth in sociocultural topics, particularly identity and equity in science education. Cognate disciplines influence the field of science education research. Cognitive psychology, sociology, and philosophy of science have significantly influenced topic emergence. Additionally, important policy reports correlate with topic spikes. Odden et al.'s (2021) review highlights the dynamic, evolving nature of science education research and its responsiveness to societal and disciplinary shifts based on publication trends in one journal.

A more expansive review is offered by Tosun (2024) who analyses 13,242 articles drawn from 14 SSCI-indexed science education journals (1982–2021) using bibliometric methods. Tosun's analysis focuses on trends in keywords, abstracts, authorship, citations, and international collaborations. In terms of general publication trends, a significant growth in publication volume since 2007 was observed, peaking in 2020. Funding support is identified as a key driver of publication growth, with 17.9% of articles having funding support. Top keywords include science education, STEM/STEM education, nature of science (NOS), assessment, professional development, scientific literacy, argumentation, gender, and conceptual change. Core topics include Nature of Science (NOS), STEM and attitudes, whereas topics in which there is growing research interest include teacher education and professional development. Most research originates from the U.S. (43.5%), with the UK, Australia, Turkey, Canada, Israel, Germany, and Taiwan also making significant contributions. The U.S. also demonstrates the strongest international collaboration network.

Research topics vary by country, with U.S. research focused on NOS, STEM, assessment, and professional development (Tosun, 2024). Research on scientific literacy, attitudes towards science, and gender originated primarily from the UK, Australia, and Canada, whereas NOS, conceptual change, and argumentation feature in Turkish research, and STEM, argumentation, and conceptual change feature in research from Taiwan. Like Odden et al. (2021), Tosun (2024) examines temporal shifts in research foci. Research from 1982 to 2006 focuses on peer review, science teaching and learning, and assessment; 2007–2016 reveals an emphasis on NOS, professional development, and conceptual change; 2017–2021 sees the rise of STEM-related research, as well as argumentation, self-efficacy, and motivation. A decline in publications on conceptual change (cf. Odden et al., 2021) and attitudes is observed in recent years. Tosun's (2024) review highlights the growing internationalization and diversification of the field of science education research.

Most recently, Lin et al. (2025) undertook a systematic content analysis of 1,142 publications in three leading international science education journals from 2018 to 2022: *Science Education*, *Journal of Research in Science Teaching*, and *International Journal of Science Education (Part A)*. Lin et al. (2025) finds that the most studied topics include learning contexts and characteristics (28.6%), teaching practices and teacher cognition (17.7%), and teacher education (12.2%). An emerging interest in cultural, social, and gender issues research (11.7%) is also identified. Empirical studies dominate the types of papers published during the reference period (90.6%), with small numbers of position papers (3.2%), literature reviews (3.2%), and theoretical papers (2.2%) also present. Replicating the country trend that Tosun (2024) reports, Lin et al.'s review finds the U.S. to be the top contributor of research to these journals, particularly *Science Education* and *Journal of Research in Science Teaching*; however, non-English-speaking countries, particularly Germany, China, and Taiwan, are well represented, most notably in the *International Journal of Science Education (Part A).* Lin et al.'s (2025) analysis reveals a shift from research regarding conceptual change and science content knowledge to identity, equity, and sociocultural perspectives, as well as increasing interdisciplinarity and attention to contemporary problems and policy relevance.

Despite the absence of emotion in the three reviews above, science education turned its attention to the study of emotions for more than 15 years as previously noted. A systematic review of the literature is both timely and necessary to chart the evolution of emotions research in science teacher education. Our review is warranted to address limitations of recent trend studies in science education research that failed to report on this sub-field of inquiry (cf. Lin et al., 2025; Odden, et al., 2021; Tosun, 2024). To this end, we analyze publication trends, the field and scope of topics, theories of emotion, research methodologies, and empirical outcomes. This review is also timely in view of unprecedented challenges to the teaching profession. Understanding emotional trajectories and challenges across the career span of science teachers can offer important insights for designing supportive teacher education programs and professional development initiatives – work that has become critically important in the context of global concerns regarding teacher shortages and retention (UNESCO, 2023). In this chapter, we offer an overview of the development of ideas about the emotions in the discipline of sociology, philosophy, and psychology, before presenting a definition adopted throughout this volume. We then define other key terms including emotion, learning, and teacher education, before presenting the questions guiding our systematic review. An overview of the book is provided at the end of the chapter to guide readers.

1.1 Defining Emotion, Learning, and Science Teacher Education

1.1.1 The Emotions-Brief History and Definition

As a field of inquiry in Education, the study of emotion predominantly owes its conceptual roots to the disciplines of sociology, philosophy, and psychology. From a scholarly perspective, authors define emotions differently in alignment with these disciplinary traditions (cf. Bellocchi, 2019). We initially outline key conceptual foundations drawing on historical accounts from Scarantino (2024), and Gardiner et al. (1937) and Dixon's (2012) histories of (Western) theories of feelings and emotion.

Historical roots of Western concepts run deep, and we liberally skip key contributors and years of progress to highlight some influential scholars.

A history of emotions outlines how the English term *emotion,* was borrowed from the French *émotion,* where, in the eighteenth century it carried the meaning of bodily stirrings and associated mental feelings which eventually came to be psychological states in the mid-nineteenth century in psychology and philosophy (Dixon, 2012). Before these times, the terms *passions* and *affections* were used by key scholars (e.g., the Stoics, Aristotle, St Thomas Aquinas, St Augustine) respectively, who were interested in mental states that are now classed as emotions.

In philosophy, three overlapping traditions relating to the emotions have been identified: 1) The *feeling tradition* holds that emotions are feelings of a distinctive type, where a feeling is a conscious experience or a sensation or a subjective quality or a quale or a what-it-is-likeness…"; 2) "The *motivational tradition* holds that emotions are motivational states of a distinctive type, or patterns of behavior of a distinctive type."; and 3) "The *evaluative tradition* holds that emotions are essentially distinguished from one another by the evaluations they involve, where an evaluation is a cognition, or an interpretation, or a judgment, or a thought, or a construal, or some other kind of mental representation of the eliciting circumstances." (Scarantino, 2016; p. 7, 15, 24). Elements of these overarching traditions are clear in historical accounts of affections, feelings, sentiments, and passions, among the array of terms that formed the focus of scholarship before *emotion* came to describe sometimes similar ideas.

Given the strong association that would arise in the 19th Century between emotion and the mind, Hippocrates is a key figure who connected the passions to the brain (Gardiner et al., 1937). Such associations marked a deviation from earlier Greek thought about *feelings* in terms of pleasure and pain, which was explained predominantly through corporeal elements. For example, from Empedocles to Anaxagoras there was recourse to explain feelings of pleasure and pain as arising from disturbance in the body such as the blood around the heart.

Leaping forward to the period of the Stoics (300), a focus on the passions arises, and these passions are seen as disturbances of the soul—excessive and overpowering impulses that are irrational and insubordinate to reason and work against nature (Gardiner et al., 1937). With the Stoics, we find a strong indication that emotions (i.e., the passions) are in opposition to and problematic for reason or rationality; a recurring theme through the development of the concept of emotion that is still with us today. Fundamental ideas, captured by Aristotle, that are still evidenced today in different theories of emotions, are that emotions can be treated as *natural kind concepts* (e.g., phenomena such as rocks or atoms) or *normative kind concepts* (Scarantino, 2024). Treated as natural kinds, inquiry into emotions looks to find their properties, like the natural sciences might seek to obtain the structure and properties of rocks. Inquiry following the normative kind assumption looks to arrive at concepts through the study of everyday meanings people assign to the passions. Scarantino (2024) notes that some contemporary emotion theories carry both assumptions. Aristotle (c. 384–322) also promoted the idea that the passions were something that affects us, rather than being something we do, and subsequently Plato (c. 427–347) offers that passions oppose reason.

In contrast to the Stoics' disdain for the emotions, St Augustine (354–430) and St Aquinas (1224/5–1274), founders of Christian psychology, critically defined distinctions between "troubling and potentially sinful movements of the soul" (i.e., appetites, lust, desires and passions), from the affections, from the Greek *pathē*, which represent, virtuous, godly, and enlightened acts (Dixon, 2012; p. 60). Passions were understood as passion of the body, brought about by activity of the lower part of the will from self-initiated activity. Affections corresponded to the higher will, hence their associations with godliness. Such distinctions between passions and affections marked a difference in conceptualizations of emotions with later ideas arising from René Descartes (1596–1650). Influencing thought on the passions for 150 years, Descartes associated passions with perceptions, which were bodily phenomena not associated with the will. Descartes, like Aquinas, attributes the causes of emotions to things external to a person, while concurrently representing a state of mind (Gardiner et al., 1937; Shapiro, 2024). According to Dixon (2012), Descartes's model removed the role of the will, and therefore human agency, from the passions, which increasingly positioned the body and became associated as their locus.

Moving to the 1800s, lectures delivered by Thomas Brown, a Doctor of Medicine and Professor of Moral Philosophy, posthumously published in 1820 would lead to a differentiation of two broad categories of mental states: internal and external affections. External affections correspond with what we call the senses today (i.e., touch, smell, etc.). Internal affections were sub-divided into thoughts and emotions (Dixon, 2012). Brown's theory was impactful in displacing terms such as passions and affections, with *emotion,* and brought together secular psychological ideas that had begun to shift away from Christian psychology. Leaping forth to 1850, the term emotion engulfs ideas such as passion, affectation, and sentiment in most English psychological lexicon. At that time, David Hume was a key figure who made early and frequent use of the word emotion in his writing and applied it in a way that is consistent with contemporary usage in psychology.

Although Christian psychology *maintained* connections between emotion and thought, it was the influence of scholars in the physicalist tradition (Alexander Bain 1818–1903, Herbert Spencer 1820–1903, Charles Darwin 1809–1882, Thomas Huxley 1825–1895) that led to the enduring separation of constructs. Here, Darwin is credited with the greatest attention, but the distinction was most pronounced in Bain's work. These works, in turn, would shape William James' (1842–1910) perspective, which established the external, involuntary, bodily nature of emotion (Dixon, 2012). In contrast to Christian psychology which maintained a connection between cognition and emotion, their separation, Dixon argues, was constructed by the physicalists.

As these brief examples recount, Scottish philosophers and thinkers were instrumental in consolidating the term and meaning of *emotion* for secular psychology. In the nineteenth century, their ideas would become imports in the U.S. and England. Through recounting this brief history, as mapped by Dixon (2012), we find an idea about the emotions that remain entrenched in some theories and lay perspectives today: the differentiation of emotion from thought. By the mid-nineteenth century, as natural science and philosophy establish their divorce, the division, present in early Greek thought, between emotion as natural kind versus normative kind grows (Scarantino, 2024). We end our brief overview of the historical accounts of emotion at this point, acknowledging that we provide merely a taste of this long and complex history. Scarantino's (2024) edited volume offers a comprehensive and detailed account of the history of emotion, which interested readers should consult. As we outline below some of the basic disciplinary approaches in psychology and sociology, that are more recent, a sense of recent history of theories, and themes from the past, will become apparent.

Sociological focus on emotion is as old as the discipline itself, which places development of sociological perspectives on emotion in a similar timeframe as the development of psychology as a discipline (cf. Stets & Turner, 2008). Over time, five main sociological traditions evolved including dramaturgical, structural, symbolic interactionist, ritual approaches, and exchange approaches. These early developments were followed by the establishment in the 1970s of the *sociology of emotion* as a consolidated sub-discipline within sociology (Olson et al., 2017). The end of the decade saw the publication of a landmark work by Arlie Hochschild on gendered emotion management that led more sociologists to engage in the study of emotion (Lively, 2024). Accepting or taking psychological perspectives on emotion for granted, sociologists attend to social contexts in which emotions are embedded. Emotions are constructed as cultural and structural phenomena, whereby culture presents a normative approach to how emotions are defined, what cultural norms drive understandings and displays of emotions, whereas structure attends to relations and organizational dimensions, such as hierarchical stratification in society (Lively, 2024). Social structures present people with affordances and constraints, and cultural norms lead to identifiably different ways in which structure operates in each societal context. Cutting across nation states, institutions, and small groups, or macro, meso, and micro levels of social life, the sociology of emotion has expansive ambitions (Lively, 2024; Turner, 2007). The sociology of emotion, Lively (2024) argues, has also shaped other fields such as cultural studies (cf. Ahmed, 2010) although without acknowledgement.

When emotion theories from the disciplines share similarity, it is to be found commonly in the identification of emotional components. For example,

> [c]omponents [of emotions] include cognitive appraisals of objects and events, subjective feelings (emotional experience), proclivity to take an action (action tendency), changes in facial configurations and vocalizations (motor component), and physiological changes occurring within the body. (Bellocchi, 2019, p. 2)

Influential examples are found in psychology and sociology (cf. Scherer, 2005; Thoits, 1989) and corroborated by recent cross-cultural comparisons in philosophy (Scarantino, 2024). A key controversy in emotion theory is the extent to which there are universally shared emotions common to humans from all cultures (Ekman, 2007; Feldman-Barrett, 2017). Drawing on Charles Darwin's work on studies of facial expressions in humans and non-human species, Ekman (2007) conducted some of the most influential studies of emotion in psychology during the twentieth century. He and his colleagues studied six facial expressions commonly associated with emotions (anger, fear, happiness, disgust, and surprise) in cross-cultural contexts. Outcomes from a range of such studies led to the argument that similarity in facial expressions associated with the respective emotion labels provides evidence that these emotions are foundational for the human species, and not culturally dependent. Subsequent failed attempts to reproduce the findings of Ekman's studies led Feldman-Barrett (2017), among others, to critique the basic emotions views. As a result of these critiques, Feldman-Barret advances the view that the brain constructs emotions (i.e., a constructionist view) in the moment and that cultural labels attached to these emotions are indicative of cultural differences rather than universality of emotions. Using examples of emotion labels from different cultures, Feldman-Barret argues that basic emotions view is simplistic at best. The view that emotions are socially constructed is also found in sociology (cf. Lively, 2024). Although the focus is less on neurology and more on social and cultural practices that shape how we define, think, and label emotions, and how these elements shape individuals and collectives.

The disciplinary traditions discussed so far reflect Western philosophical foundations, and there is relatively less English-language scholarship representing Eastern philosophy. Examples of Eastern perspectives reveal that the term emotion is not a collective noun, like it is in English, for other labels that described specific emotions like anger, joy, happiness (Dalai Lama & Thubten Chodron, 2017; de Silva, 2011; Sonam Choegyal, 2019; Sogyal Rinpoche, 1992). To provide examples from Buddhism, specific labels that parallel English emotion labels exist: *gawa [joy], chowa [sad], dro [fear], tsipza [angry], khamlok [disgust]*. In contrast to these analogous terms from Western and Eastern perspectives, the closest term to the English collective noun *emotion* in Bhuddism is *sem*, which refers to the ordinary mind. Sogyal Rinpoche (1992) exemplifies the connection between sem and the English-language equivalent of specific emotions as follows:

> even to think of the possibility of enlightenment, one looks at what composes our ordinary mind- anger, greed, jealousy, spite, cruelty, lust, fear, anxiety, and turmoil- would undermine forever any hope of achieving it, if we had not been told about the nature of mind, and the possibility of coming to realize that nature beyond all doubt (p. 54).

In contrast to sem, which is influenced by anger, greed, jealousy, spite, cruelty, lust, fear, anxiety, and turmoil, Buddhism differentiates *rigpa*—the awakened mind of awareness (Sogyal Rinpoche, 1992). Bhuddism seems to focus on the effects of specific emotions in leading to either an unenlightened or enlightened mental state. Further to this, we refer readers to chapters in Scarantino's (2024) collection, that address ancient and classical Indian and early and medieval Chinese theories.

Although some scholars have challenged notions that emotion labels represent universal psychological states, as discussed previously, analysis of emotion concepts across cultures reveals that there are shared elements. Ancient thought from Greece, Rome, India, and China reveals the following cross-cultural similarities in defining emotions:

- Experiences, feelings, qualia- often considered as valenced and bodily
- Action tendencies, desires, or endeavors

- Appraisal, evaluations, and cognitions (Scarantino, 2020).

In this respect, these shared elements reflect what the componential theories of emotion in psychology and sociology use to define emotions (cf. Scherer, 2005; Thoits, 1989).

Broadly speaking, science education research draws theoretical and conceptual impetus from sociology, psychology, philosophy, feminist perspectives, cultural theory, historical inquiry, and anthropology (cf. Bellocchi, 2019; Pekrun & Linnenbrink-Garcia, 2014). Due to this conceptual diversity, there is no way of presenting a single definition of emotion that will satisfy all the usage we encounter in preservice and inservice science education research through our review (cf. Bellocchi & Amat, 2022). Next, we present one of the more widely shared theorizations but urge readers to engage with the specific theory or perspective outlined by authors of the original articles included in Chaps. 4–9.

We define emotions following a componential theory because this provides the greatest theoretical latitude for the reasons outlined below. Sociologist, Peggy Thoits (1989), defines emotions as having four components including linguistic (emotion) labels, expressive gestures, situational cues, and physiological changes. In contrast, feelings are generalized physical states (like hunger, fatigue) and affects are positive and negative appraisals of experiences (e.g., expressing dislike of something; Thoits, 1989). In a related way, Bendelow and Williams (1998) identify physiological, interactional, social, and cultural dimensions to emotions. This approach to defining emotions based on their components is also evident in a dominant psychological perspective captured by Klaus Scherer (2005). In Scherer's (2005) definition, emotions involve interrelated changes in all or most of the five organismic subsystems: support; information processing; monitoring; executive function; and action. Changes in these subsystems occur in response to an environmental or internal stimulus event that is of considerable concern to the organism. The five subsystems depend on changes in the central nervous system, neuroendocrine system, autonomic nervous system, and somatic nervous system. Componential theories of emotions have had a strong influence in education research where subjective feeling, a motor response, physiological changes, actions tendencies, and appraisals are identified as key components of emotions when a range of theories, predominantly psychological ones, are coalesced (Pekrun & Linnenbrink-Garcia, 2014).

Similar ideas are found in psychodynamic perspectives where the emotions are defined as physiological and behavioral reactions that are involved in cognitive and evaluative processes (see Zembylas, 2007 for a summary). Such evaluations prompt us to feel and act, reflecting the subjective feeling and action tendencies described in the theories discussed above.

1.1.2 Learning

In this book, *learning* refers to formal or informal situations in which science teachers acquire knowledge and experience about teaching science. This includes learning about curriculum, pedagogy, assessment, and reflecting on teacher learning experiences. Learning can take place in formal or informal settings. Like emotions, researchers in our field will either choose a specific definition of learning from the vast array on offer, or, more commonly, they do not define the term, leaving implicit any assumptions about the way in which preservice and inservice science teachers learn. For these reasons, we offer Illeris's (2018) summary of the elements of learning that are found across a range of dominant learning theories, which apply across the life course because his approach is sufficiently expansive to cover elements encountered in preservice and inservice science teacher education research. Through his synthesis, Illeris identifies three dimensions of learning processes: a *content dimension*—mostly but not always cognitive; an *incentive dimension* which is mainly affective (motivation, interest, engagement); and an *interactional dimension* covering social and material aspects of the environment at local and global levels. Learning occurs through an individuals' impressions, which are formed through interactions between learning and the social/material environment, and through a process of acquisition in which an individual assesses, elaborates, and takes in their impressions.

Illeris' synthesis of learning theories encompasses emotions and affect (i.e., motivation, interest, and engagement). As he notes in his historical review of learning theories, it took time before emotion and affect became recognized as important for learning.

1.1.3 Science Teacher Education

Throughout this book, 'preservice science teacher' refers to an individual who is undertaking university study at the undergraduate or postgraduate level to become a qualified teacher of science at any level of schooling (i.e., early childhood, elementary, secondary), but has not yet started teaching professionally (Peterson, et al., 2009). By contrast, an 'inservice science teacher' is a professionally qualified teacher working in educational settings such as schools and early learning centers who teaches science to children and young people. We refer to all formal education for preservice and inservice science teachers, including university study and professional development, as 'science teacher education.' In the context of preservice teachers, specifically, university-based teacher education is also referred to as 'initial teacher education' (ITE) (Ell et al., 2019).

Having offered an introduction to the goals of the book and clarified key terms, we present the research questions answered by our systematic review before presenting the review methodology in Chap. 2.

1.2 Research Questions Addressed in This Book

Considering the growing body of research on emotion in science teacher education, we are well placed to ask and answer more specific questions about the status of knowledge in the field:

1. What topics associated with emotions have researchers addressed in the context of preservice and inservice science education?
2. What are the main outcomes from research on emotions in preservice and inservice science education?
3. How are emotions theorized in research on emotion in preservice and inservice science teaching?
4. What methods have been used in emotions research in preservice and inservice science education?

Most recently, Bellocchi and Amat (2022) offer a partial answer to the first question by reviewing research on the personal and professional emotional growth of science teachers through a review of studies of emotion in preservice and inservice science teaching and learning. In this volume, we offer a more comprehensive review of studies. Herein, we use the term *science teacher education* to encompass preservice and inservice science teaching and learning.

As noted by Bellocchi and Amat (2022), early work like Zembylas' (2001), introduces the idea of researching emotions in science education by drawing on studies of *affect* and studies outside of science education to build key arguments about the importance of investigating the role that emotions play in science teacher education. Thanks to such pioneering work, slow growth in publications reporting science teacher emotions, as opposed to the more general construct of affect, began to emerge, often alongside considerations of student emotions rather than being separate foci (cf. Davis & Bellocchi, 2018; Sinatra et al., 2014).

There is currently no comprehensive review that outlines the landscape of existing research on emotions in science teacher education and summarizes key theoretical trends and empirical outcomes. To synthesize existing work and to offer impetus for future research, this book addresses the lack of a systematic review of the field. We also present a contemporary summary of empirical outcomes from research on emotions in science teacher education.

1.3 Overview of the Book

In Chap. 2 we discuss the systematic review methodology applied for searching and screening articles for inclusion, which yielded 93 articles. An overview of publication trends is the topic of Chap. 3. Here, we present the landscape of the field of research by outlining the volume of publications over time, author networks, disciplinary origins of emotion theories, study designs, schooling levels, and sample countries.

Chap. 3 ends with the outcomes of our thematic analysis of topics that form the focus of science teacher emotions research reporting emotions or emotional outcomes. The six themes resulting from our analysis become the organizing ideas for our systematic review of empirical outcomes of research in Chaps. 4–9. The chapters begin with a general introduction to the included papers, before providing a review of dominant theoretical trends in the theme. This is followed by detailed discussion of included papers, and the chapters end with summaries that draw together the main ideas for each theme. Chap. 4 reports empirical outcomes of the *Classroom emotional climate* theme, which had the fewest publications of all themes. Practical aspects of science are captured in Chap. 5, which focuses on the theme *Science inquiry, laboratory, and practical work*. *Controversial issues* is the theme name at the center of Chap. 6. Readers will be familiar with one topic in that theme; socio-scientific issues. In Chap. 7, *Technology* is the theme, whereas Chap. 8 focuses on *Self-efficacy and Identity*, two very closely related topics in the included studies. The theme with the most numerous included articles ends our thematic chapters; (*Science teachers' work*; Chap. 9). Presenting our systematic review in book format allows us to discuss each included study in-depth in our thematic chapters. We conclude the book in Chap. 10 by looking across our review findings and proposing directions that future research on science teacher emotions can take.

1.4 Conclusion

This chapter has provided the rationale and scope for this volume that reports on our systematic review of preservice and inservice science teachers' emotions. Key terminology including emotions, learning, and science teacher education were defined and a brief history of emotion scholarship was canvassed. Research questions that guided the review were presented before an overview of the volume was outlined.

References

NB Articles included in the systematic review are indicated using an asterisk (*)

Ahmed, S. (2010). *The promise of happiness*. Duke University Press.

Bellocchi, A. (2019). Emotions and teacher education. In G. Noblit (Ed.), *The Oxford research encyclopaedia of education*. Oxford University Press. https://doi.org/10.1093/acrefore/9780190264093.013.773

Bellocchi, A., & Amat, A. (2022). Emotion and science teacher education. In J. A. Luft, & M. G. Jones, (Eds.), *Handbook of research on science teacher education* (pp. 426–438). Routledge. https://doi.org/10.4324/9781003098478-38

*Bellocchi, A., Mills, K., Olson, R., Patulny, R., & McKenzie, J. (2018b). Emotion work at the frontline of STEM teaching. In L. Bryan & K. Tobin (Eds.), *Critical issues and bold visions for science education: The road ahead* (pp. 247–264). Brill. https://doi.org/10.1163/9789004389663_014

Bendelow, G., & Williams, S.J. (Eds.). (1998). *Emotions in social life: Critical themes and contemporary issues* (1st ed.). Routledge. https://doi.org/10.4324/9780203437452

Dalai Lama & Thubten Chodron. (2017). *Approaching the Buddhist path.* https://ebookcentral.proquest.com

Davis, J. P., & Bellocchi, A. (2018). Emotions in learning science. In S. M. Ritchie & K. G. Tobin (Eds.), *Eventful learning: Learner emotions* (pp. 9–30). Brill. https://doi.org/10.1163/9789004377912_002

de Silva, P. (2011). Thinking and feeling: A Buddhist perspective. *Sophia, 50*(2), 253–263. https://doi.org/10.1007/s11841-011-0248-2

Dixon, T. (2012). *From passions to emotions: The creation of a secular psychological category.* Cambridge University Press.

Ell, F., Simpson, A., Mayer, D., McLean Davies, L., Clinton, J., & Dawson, G. (2019). Conceptualizing the impact of initial teacher education. *The Australian Educational Researcher, 46*, 177–200.

Ekman, P. (2007). *Emotions revealed*. St Martin's Griffin.

Feldman-Barrett, L. (2017). *How emotions are made: The secret life of the brain.* Macmillan.

Gardiner, N. H., Clark Metcalf, R., & Beebe-Centre, J. G. (1937). *Feeling and emotion: A history of theories.* American Book Publishing. https://doi.org/10.1037/10763-000

Illeris, K. (2018). An overview of the history of learning theory. *European Journal of Education, 53*, 86–101. https://doi.org/10.1111/ejed.12265

Lin, T.-J., Lin, T.-C., Potvin, P., & Tsai, C.-C. (2025). Research trends in science education from 2018 to 2022: A systematic content analysis of publications in selected journals. *International Journal of Science Education, 47*(4), 510–538. https://doi.org/10.1080/09500693.2024.2340811

Lively, K. J. (2024). An overview of contemporary theories of emotions in sociology. In A. Scarantino (Ed.), *Emotion theory: The Routledge comprehensive guide: Volume 1: History, contemporary theories, and key elements* (pp. 287–309). Routledge. https://doi.org/10.4324/9781315559940-17

Odden, T. O., B., Marin. A., & Rudolph, J. L. (2021). How has *Science Education* changed over the last 100 years? An analysis using natural language processing. *Science Education, 105*, 653–680. https://doi.org/10.1002/sce.21623

Olson, R. E., McKenzie, J. J., & Patulny, R. (2017). The sociology of emotions: A meta-reflexive review of a theoretical tradition in flux. *Journal of Sociology, 53*(4), 800–818. https://doi.org/10.1177/1440783317744112

Pekrun, R., & Linnenbrink-Garcia, L. (Eds.). (2014). *International handbook of emotions in education.* Routledge.

Peterson, P., Baker, E., & McGaw, B. (2009). *International encyclopedia of education.* Elsevier.

Ritchie, S. M., & Tobin, K. G. (Eds.). (2018). *Eventful learning: Learner emotions.* Brill.

Scarantino, A. (2016). The philosophy of emotions and is impact on affective science. In L. F., Barrett, M., Lewis, & J. M. Haviland-Jones, (Eds.), *Handbook of emotions* (4th ed., pp. 3–48). The Guilford Press.

Scarantino, A. (2020, September, 25). Some recurrent patterns in the history of emotion theory. International Society for Research on Emotion Webinar Series 2020. https://www.isre.org/news/526934/Emotion-Webinar---Lecture-7.htm

Scarantino, A. (2024). Introduction to volume 1: The value of history, a wealth of theoretical options, and the elements of emotion theory. In A. Scarantino (Ed.), *Emotion theory: The Routledge comprehensive guide: Volume 1: History, contemporary theories, and key elements* (pp. 1–13). Routledge. https://doi.org/10.4324/9781315559940-1

Scherer, K. R. (2005). What are emotions? And how can they be measured? *Social Science Information, 44*(4), 695–729.

Shapiro, L. (2024). Emotion theory in the 17th century. In A. Scarantino (Ed.), *Emotion theory: The Routledge comprehensive guide: Volume 1: History, contemporary theories, and key elements* (pp. 110–128). Routledge. https://doi.org/10.4324/9781315559940-8

Sinatra, G. M., Broughton, S. H., & Lombardi, D. (2014). Emotions in science education. In R. Pekrun & L. Linnenbrink-Garcia (Eds.), *International handbook of emotions in education* (pp. 415–436). Routledge.

Sogyal Rinpoche (1992). *The Tibetan book of living and dying*. Harper Collins.

Sonam Choegyal. (2019). *A study of Bhutanese 9th grade physics students' emotions and social bonds during transformative pedagogy* (Unpublished doctoral dissertation). Queensland University of Technology.

Stets, J. E., & Turner, J. H. (2008). The sociology of emotions. In M. Lewis, J. M. Haviland-Jones, & L. Feldman Barrett (Eds.), *Handbook of emotions* (3rd ed., pp. 32–44). Guilford.

Thoits, P. A. (1989). The sociology of emotions. *Annual Review of Sociology, 15*, 317–342.

Tosun, C. (2024). Analysis of the last 40 years of science education research via bibliometric methods. *Science & Education, 33*, 451–480. https://doi.org/10.1007/s11191-022-00400-9

Turner, J. H. (2007). *Human emotions: A sociological theory*. Routledge.

United Nations Educational, Scientific and Cultural Organization. (2023). *Global report on teachers: Addressing teacher shortages.* https://teachertaskforce.org/what-we-do/Knowledge-production-and-dissemination/global-report-teachers

*Zembylas, M. (2001). A paralogical affirmation of emotion's discourse in science teaching. In A. Barton & M. Osborne (Eds.), *Teaching science in diverse settings: Marginalized discourses and classroom practice* (pp. 99–128). Peter Lang.

Zembylas, M. (2007). Emotional ecology: The intersection of emotional knowledge and pedagogical content knowledge in teaching. *Teaching and Teacher Education, 23*, 355–367. https://doi.org/10.1016/j.tate.2006.12.002

Chapter 2
Systematic Review Method

This systematic review follows the Preferred Reporting Items for Systematic Reviews and Meta-Analyses (PRISMA) guidelines (Page et al., 2021), with adaptations appropriate for the educational context of our review. PRISMA provides a structured framework for conducting and reporting systematic reviews to ensure transparency, reproducibility, and comprehensiveness. Our aims in presenting this review are to provide an overview of the scope and trends of research on the emotions in science teacher education, outline main outcomes related to the emotions and science teacher education, and use these two dimensions to suggest fruitful future research directions (cf. Grant & Booth, 2009).

In the absence of existing reviews, based on a scoping search, the research questions presented in Chap. 1 are appropriate for charting the field and making recommendations for future research. The questions provide sufficient breadth to identify the nature of research and how it has been conducted, and to identify research outcomes specific to science teacher education associated with different stages of schooling (elementary, secondary, tertiary), such that the answers will encompass university experiences, practicum, and school-based teacher learning such as professional development. The nature of the questions also focuses attention on empirical research.[1] Given that empirical research in this field is still emerging, such broad questions are feasible and will provide sufficient overview of the status of the topic in the field at present.

[1] Disambiguation: Empirical research is not restricted to research conducted within the assumptions of British Empiricism, which can be seen as a research paradigm.

[3]Web of Science produced the same results as other databases with no unique outcomes. For this reason, it was no longer used.

A. Bellocchi et al., *Emotion Research in Science Teacher Education*, Sociocultural Explorations of Science Education 32, https://doi.org/10.1007/978-3-032-07711-0_2

2.1 Literature Search Process

Following PRISMA guidelines, we designed and implemented a comprehensive search strategy to identify all relevant literature on emotions in science teacher education. We conducted a systematic review, mapping the range and nature of concepts and approaches evident in empirical literature on emotions in science teacher education (cf. Grant & Booth, 2009). Focusing on empirical articles allowed us to identify key information from a field of research where there is wide diversity in the kinds of research designs used to study emotions.

Our search was conducted using the earliest date in each database (i.e., no start date specified) with an end date of 7th April, 2025. All databases[3] available to the researchers were accessed and searched because different databases can yield different search results. ERIC is the largest database of educational research, and to this we added databases from psychology and the social sciences to capture the widest range of publications that may not appear in education books or journals. We searched the following 9 databases:

1. ERIC
2. Education Source
3. APAPsychInfo
4. APAPsych Articles
5. APAPsych Books
6. Academic Search Elite
7. eBook Collection
8. ProQuest
9. Scopus

Drawing on 15 years' research experience in researching emotions in science education, we **delimited the searches to title, abstract, and keywords of peer-reviewed articles**. There is variation across databases in terminology, such that when *keywords* were not an option, *subject terms* or *all subjects* were selected as the equivalent option. In all databases, searches were screened for peer reviewed articles written in English.

2.2 Search Term Development

We used the terms *emotion, affective,* and *feeling* initially to identify articles relevant for our focus. This choice is grounded in the interchangeable ways in which some authors use the terms *affect, feelings,* and *emotion.* The truncation *emoti** was used to capture derivatives including emotion, emotive, emotional to produce the following search string (emoti* OR affective OR feeling*). This string was combined with the following strings related to science teachers using the AND function:

String 1
Preservice/Inservice Education String

("preservice science teach*" OR "pre-service science teach*" OR "preservice science stud*" OR "pre-service science stud*" OR "inservice science teach*" OR "in-service science teach*" OR "beginning science teach*" OR "science teacher train*")

String 2
Prospective Science Teachers Emotions

("prospective science teachers" OR "prospective physics teachers" OR "prospective chemistry teachers" OR "prospective earth science teachers" OR "prospective biology teachers")

String 3
Beginning Science Teaching String

("beginning physics teach*" OR "beginning chemistry teach*" OR "beginning biology teach*" OR "beginning earth science teach*" OR "beginning* science teach*" OR "beginning engineering teach*")

String 4
Engineering Education String

("preservice engineering education" OR "pre-service engineering education" OR "in-service engineering education" OR "inservice engineering education") String 5

Professional Development/Professional Learning String

("science teach*" OR "science educ*" OR "science learn*") AND ("professional development" OR "professional learning" OR "continuing education" OR "ongoing educat*" OR "on-going educat*")

2.3 Additional Search Strategies

Books and book chapters were initially identified through the above database searches and then extended through the authors' knowledge of the field, and author contacts. We accessed several edited collections and initially identified 16 book chapters that were potentially relevant for the review as shown in Table 2.1. During initial screening using the inclusion criteria and a comparison of book chapters by Zembylas (2005a) with journal articles by the same author (i.e., Zembylas, 2002, 2004a, 2004b), the 4 chapters at the bottom of Table 2.1 were excluded due to high similarity with journal publications from the same author. All other chapters in Table 2.1 progressed to full-text review for final inclusion decisions, after which 10 chapters were included for meeting the inclusion criteria.

Table 2.1 Books and edited collections screened for emotion research in preservice and inservice science teacher education

Editor	Year	Title	Chapters identified for screening
Alsop, S	2005	*Beyond Cartesian dualism: Encountering affect in the teaching and learning of science*	**1.** Shapiro, B. (2005). From despair to success: A case study of support and transformation in an elementary science practicum. (pp. 161–172) **2.** Zembylas, M. (2005a). Emotions and science teaching: Present research and future agendas. (pp.123–132)
Bellocchi, A., Quigley, C. & Otrel-Cass, K	2017	*Exploring emotions, aesthetics, and wellbeing in science education research*	**3.** Hobbs, L., & Kelly, L. (2017). The heart of the educator: Aesthetic experience shaping knowledge, identity, and passion. (pp. 55–84) **4.** Cook, K. & Buck, G. (2017). Our neighbourhood: A place for heightening emotional energy in science education. (pp. 141–164) **5.** Ritchie, S.M., & Newlands, J. (2017). Emotional events in learning science. (pp. 107–121)
Bryan, L., & Tobin, K.G	2018	*Critical issues and bold visions for science education: The road ahead*	**6.** Bellocchi, A., Mills, K.A., Olson, R., Patulny, R., & McKenzie, J. (2018b). Emotion work at the frontline of STEM teaching: Burning out and turning away. (pp. 247–264)
Pekrun, R., & Linnenbrink-Garcia, L	2014	*International handbook of emotions in education*	**7.** Sinatra, G., Broughton, S.H. & Lombardi, D. (2014). Emotions in science education. (pp.415–436)
Ritchie, S. M. & Tobin, K	2018	*Eventful learning: Learner emotions*	**8.** Bellocchi, A., Davis, J.P., & King, D. (2018a). Science demonstrations as mediators of emotional experiences. (pp. 57–86)
Schutz, P., & Zembylas, M	2009	*Advances in teacher emotion research*	**9.** Bullough, R. V. (2009). Seeking Eudaimonia: the emotions in learning to teach and mentor **10.** Cross, D.I., & Hong, J.Y. (2009). Beliefs and professional identity: Critical constructs in examining the impact of reform on the emotional experiences of teachers **11.** Rosiek, J., & Beghetto, R.A. (2009). Emotional scaffolding: The emotional and imaginative dimensions of teaching and learning. (pp. 175–194)

(continued)

Table 2.1 (continued)

Editor	Year	Title	Chapters identified for screening
Varelas, M	2012	*Identity construction and science education research: Learning, teaching, and being in multiple contexts*	**12.** Rivera Maulucci, M.S. (2012). Exploring linkages between identity and emotions in teaching for social justice in science teacher education. (pp. 123–139)
Zembylas, M	2005	*Teaching with emotion: A postmodern enactment*	**13.-16.** Chapters 4–7

In addition to database searches and contacting authors, we also engaged in a pearling process whereby the reference lists of included journal articles and book chapters identified in initial searches were searched and compared to our database to identify any new articles not found from database searches. Pearling initially yielded 36 articles, of which, after removal of duplicates from our data set and application of inclusion criteria, 9 were retained (see Fig. 2.1). As seen in Fig. 2.1, 92 documents were retained overall including 11 book chapters and 81 journal articles. The complete set of included articles is presented in Table 2.2 (see below).

2.4 Inclusion Criteria

The following inclusion criteria, based on an adaptation of PRISMA, were used to screen article title, abstracts, and full text at separate stages of the review process shown in Fig. 2.1:

Cohort: preservice or inservice science school teachers (inc. biology, chemistry, physics, earth and environmental science, general science)
Setting: Any formal (teacher education courses and professional learning courses) or informal setting (reports of incidental teacher learning) where learning about science teaching takes place
Phenomenon: Study addressing emotions, NOT affect or feelings (cf. Thoits, 1989).
Study type: Scholarly peer reviewed empirical articles based on any research design
Outcomes: Key outcomes related to science teacher emotions are reported (not school students' emotions).

At the title and abstract screening stage, any publication that could not be included/excluded immediately based on the criteria was progressed to full-text screening.

The 92 studies indicated in Fig. 2.1 are summarized in Table 2.2 below and marked with an asterisk in the Reference section of the book. In Table 2.2, included articles are organized with book chapters appearing first, followed by journal articles.

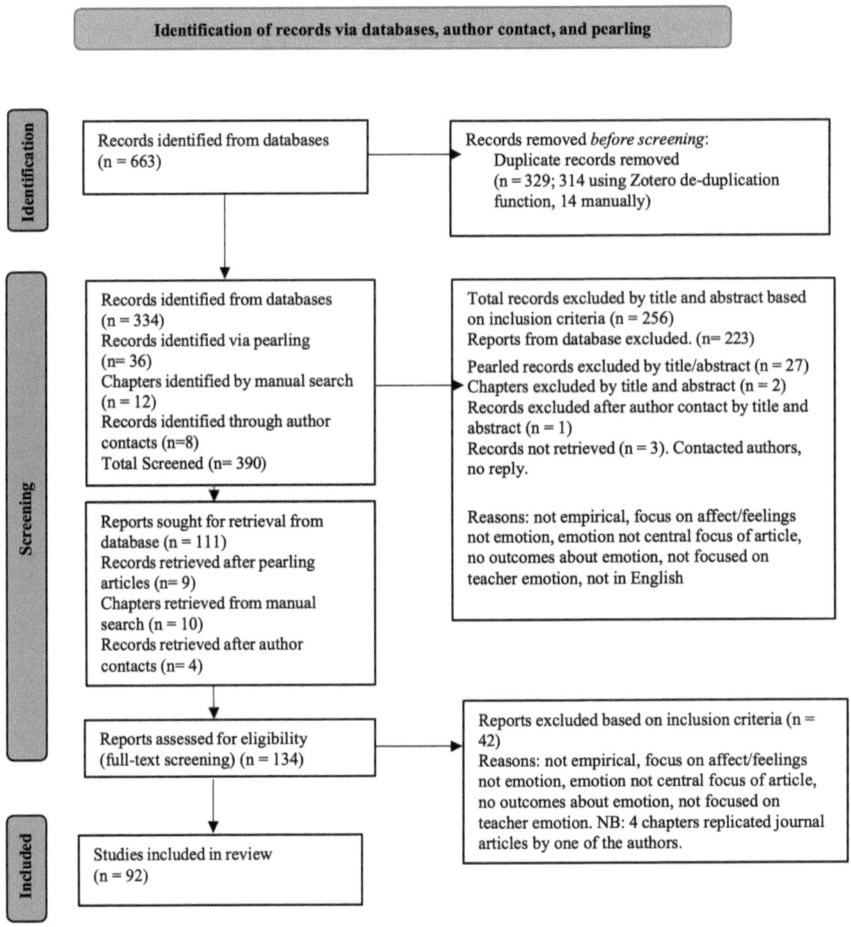

Fig. 2.1 Search results and final articles included. Adapted from Page et al. (2021)

2.5 Data Extraction and Synthesis Processes

Following PRISMA guidelines, we developed a standardized data extraction form to collect relevant information from each included study. The data extraction process focused on capturing key details about study characteristics, methodologies, theoretical frameworks, and findings related to emotions in science teacher education.

Data were extracted from each included study by the four authors independently. The following information was extracted from each study: Study characteristics (author, year, country, educational context), participant information (sample country, type of teachers - preservice/inservice), study design and methodological approach, and theoretical framework or conceptualization of emotions. The first

Table 2.2 Included articles on emotion in preservice and inservice science teacher education

#	Pub type	Author	Year	Title/brief source details	Education context	Design (0ata)	Emotion theory
1	B	Avraamidou, L.	2018	Chapter 13 Elementary science teacher identity as a lived experience: small stories in narrative analysis. In P. A. Schutz et al. (eds.), *Research on teacher identity* (pp. 145–155)	Inservice, elementary. Southern Europe	Qual (I, W)	Sociology, Psychology
2	B	Bellocchi, A., Davis, J.P., & King, D.	2018a	Science demonstrations as mediators of emotional experiences. In Tobin, K & Ritchie, S M (Eds.) *Eventful learning: Learner emotions.* Brill, The Netherlands, pp. 57–85	Preservice, high school, Australia	Qual (O, It)	Sociology
3	B	Bellocchi, A., Mills, K., Olson, R., Patulny, R., & McKenzie, J.	2018b	Emotion work at the frontline of STEM teaching: Burning out and turning away. In Bryan, L.A. & Tobin, K. (Eds.), *Critical issues and bold visions for science education: The road ahead.* Rotterdam: Sense	Inservice, elementary, high school, Australia	Qual (O)	Sociology
4	B	Bullough, R.V.	2009	Seeking eudaimonia: The emotions in learning to teach and to mentor. In Schutz, A., & Zembylas, M. (Eds.), *Advances in teacher emotion research*	Inservice, middle school, USA (inferred)	Qual (I)	Appraisal theory, Identity

(continued)

Table 2.2 (continued)

#	Pub type	Author	Year	Title/brief source details	Education context	Design (0ata)	Emotion theory
5	B	Cebesoy, Ü.B.	2025	Exploring preservice science teachers' emotive reasoning through place-based SSI Instruction: The case of gold mining. In B. Namdar & E. Karahan (Eds.). *Socioscientific issues focused teacher education* (pp.153–180). Springer	Preservice, high school, Turkey	Qual (W)	Education- SSI
6	B	Cook, K. & Buck, G.	2017	Our neighbourhood: A place for heightening emotional energy in science education. In A. Bellocchi, C. Quigley, & Otrel-Cass, K. (Eds.), *Exploring Emotions, Aesthetics and Wellbeing in Science Education Research.* (pp. 141–164). Springer	Inservice, elementary, middle school, USA	Qual (A, I, FG, O)	Sociology
7	B	Cross, D.I. & Hong, J.-Y.	2009	Beliefs and professional identity: Critical constructs in examining the impact of reform on the emotional experiences of teachers. In A. Schutz, & M. Zembyla.s (Eds.), *Advances in teacher emotion research.* Springer	Inservice, middle/ high school, USA	Qual (I)	Social constructionism

(continued)

Table 2.2 (continued)

#	Pub type	Author	Year	Title/brief source details	Education context	Design (0ata)	Emotion theory
8	B	Hobbs, L., & Kelly, L.	2017	The heart of the educator: Aesthetic experience shaping knowledge, Identity, and passion. In A. Bellocchi, C. Quigley, & K. Otrel-Cass. (Eds.), *Exploring emotions, aesthetics and wellbeing in science education research*. Springer	Pre/Inservice, elementary, Australia	Qual (I)	Aesthetics
9	B	Ritchie, S.M., & Newlands, J.	2017	Emotional events in learning science. In A. Bellocchi, C. Quigley, & K. Otrel-Cass. (Eds.), *Exploring Emotions, Aesthetics and Wellbeing in Science Education Research,* (pp. 107–121). Springer	Preservice, high school, Australia	Qual (FG, O, W)	Sociology
10	B	Rivera Maulucci, M.S.	2012	Exploring linkages between identity and emotions in teaching for social justice in science teacher education. In M. Varelas (Ed.). *Identity construction and science education research: learning, teaching, and being in multiple contexts*. (pp. 123–139). Sense Publishers	Inservice, preservice, elementary, high school, Caribbean/ Dom. Republic	Qual (A, I, O)	Discursive

(continued)

Table 2.2 (continued)

#	Pub type	Author	Year	Title/brief source details	Education context	Design (0ata)	Emotion theory
11	B	Zembylas, M.	2001	A paralogical affirmation of emotion's discourse in science teaching. In A. Barton and M. Osborne (Eds.), *Teaching science in diverse settings: Marginalized discourses and classroom practice*. (pp. 99–128). Peter Lang	Inservice, elementary, USA	Qual (I)	Multi-disciplinary
12	J	Abualrob, M.M.A. & Al-Saadi, S.H.	2025	Evaluating the role of technological pedagogical content knowledge in ICT-enhanced science education. *International Journal of Information and Education Technology*	Preservice/ inservice, elementary, Palestine	Quant (SL)	Control-value theory
13	J	Adler, I., Zion, M., & Rimerman-Shmueli, E.	2019	Fostering teachers' reflections on the dynamic characteristics of open inquiry through metacognitive prompts. *Journal of Science Teacher Education*	Inservice, kindergarten, middle school, high school, Israel	Quant + Qual (A, SL, So)	N/A
14	J	Akerson, V.L., Pongsanon, K., Weiland, I. S., & Nargund-Joshi, V.	2014	Developing a professional identity as an elementary teacher of nature of science: A self-study of becoming an elementary teacher. *International Journal of Science Education*	Inservice, elementary, USA	Qual (self-study: A, O, W)	Identity

(continued)

Table 2.2 (continued)

#	Pub type	Author	Year	Title/brief source details	Education context	Design (0ata)	Emotion theory
15	J	Alarcón-Orozco, M.M., Franco-Mariscal, A.J., Oliva, J.M., & Blanco-López, Á.	2025	Emotions experienced by preservice early childhood teachers during a training program in inquiry-based science education. *Journal of Research in Science Teaching*	Preservice, early childhood, Spain	Quant + Qual (SL, So)	N/A
16	J	Amponsah, K.D., Adu-Gyamfi, K., Awoniyi, F.C., & Commey-Mintah, P.	2024	Navigating academic performance: Unravelling the relationship between emotional intelligence, learning styles, and science and technology self-efficacy among preservice science teachers. *Heliyon*	Preservice, high school, Ghana	Quant (SL)	Emotional intelligence
17	J	Ayvaci, H.Ş., Bebek, G., & Yamaçli, S.	2023	Experiences of pre-service science teachers in "teaching practice" during the Covid-19 pandemic. *International e-Journal of Educational Studies*	Preservice, Turkey	Qual (I, O)	N/A

(continued)

Table 2.2 (continued)

#	Pub type	Author	Year	Title/brief source details	Education context	Design (0ata)	Emotion theory
18	J	Bahcivan, E.	2019	Examining the structural relations among PSTs' scientific epistemological beliefs, epistemic emotions and argumentativeness: Sample from Turkey. *International Journal of Education in Mathematics, Science and Technology*	Preservice, Turkey	Quant (SL)	Epistemic emotions, Psych
19	J	Bellocchi, A., Mills, K., & Ritchie, S.M.	2016	Emotional experiences of preservice science teachers in online learning: the formation, disruption and maintenance of social bonds. *Cultural Studies of Science Education*	Preservice, high school, Australia	Qual (W)	Sociology
20	J	Bellocchi, A., Ritchie, S.M., Tobin, K., King, D., Sandhu, M., & Henderson, S.	2014	Emotional climate and high quality learning experiences in science teacher education. *Journal of Research in Science Teaching*	Preservice, high school, Australia	Qual (SL)	Sociology, Psychology
21	J	Bellocchi, A., Ritchie, S.M., Tobin, K., Sandhu, M., & Sandhu, S.	2013	Exploring emotional climate in preservice science teacher education. *Cultural Studies of Science Education*	Preservice, high school, Australia	Qual (SL)	Sociology, Psychology
22	J	Betul Cebesoy, Ü, & Chang Rundgren, S.-N.	2023	Embracing socioscientific issues-based teaching and decision-making in teacher professional development. *Educational Review*	Preservice, elementary, Western Turkey	Qual (W)	N/A

(continued)

Table 2.2 (continued)

#	Pub type	Author	Year	Title/brief source details	Education context	Design (0ata)	Emotion theory
23	J	Borrachero, A.B., Brígido, M., Mellado, L., Costillo, E., & Mellado, V.	2014	Emotions in prospective secondary teachers when teaching science content, distinguishing by gender. *Research in Science and Technology Education*	Preservice, Spain	Quant (SL)	Multidisciplinary-Sociology, Psych, & Cognitive sci
24	J	Brick, K., Cooper, J., Mason, L., Faeflen, S., Monmia, J., & Dubinsky, J. M.	2021	Tiered neuroscience and mental health professional development in Liberia improves teacher self-efficacy, self-responsibility, and motivation. *Frontiers*	Inservice, high school, Liberia	Quant + Qual (I, SL)	Neurobiology
25	J	Brígido, M., Borrachero, A., Bermejo, M.L., & Mellado, V.	2013a	Prospective primary teachers' self-efficacy and emotions in science teaching. *European Journal of Teacher Education*	Inservice, elementary, Spain	Quant (SL)	Education
26	J	Brígido, M., Couso, D., Gutiérrez, C., & Mellado, V.	2013b	The emotions about teaching and learning science: a study of prospective primary teachers in three Spanish universities. *Journal of Baltic Science Education*	Preservice, elementary, Spain	Quant (SL)	Multidisciplinary-Sociology, Psych, & Cognitive sci

(continued)

Table 2.2 (continued)

#	Pub type	Author	Year	Title/brief source details	Education context	Design (0ata)	Emotion theory
27	J	Brígido, M., Bermejo, M. L., Conde, M. C., & Mellado, V.	2010	The emotions in teaching and learning nature sciences and physics/chemistry in pre-service primary teachers. *US-China Educational Review*	Preservice, elementary, Spain	Quant (SL)	Education
28	J	Can, H.	2015	Sources of teaching efficacy beliefs in pre-service science teachers. *Elementary Education Online*	Preservice, high school, Turkey	Qual (I)	Psychology
29	J	Cansoy, R.	2017	Teachers' professional development: The case of WhatsApp. *Journal of Education and Learning*	Inservice, high school (inferred), Turkey (inferred)	Qual (I)	N/A
30	J	Cooke, H., Campbell, T., Luehmann, A., Zhang, Y., & Scipio, D.	2025	"I sit here thinking I can do this"—Developing justice-centered ambitious science teaching identities in professional learning communities. *Journal of Research in Science Teaching*	Inservice, middle school/high school, USA	Qual (FG, O)	N/A

(continued)

Table 2.2 (continued)

#	Pub type	Author	Year	Title/brief source details	Education context	Design (0ata)	Emotion theory
31	J	Davidson, S: G., Jaber, L. Z., & Southerland, S. A.	2020	Emotions in the doing of science: Exploring epistemic affect in elementary teachers' science research experiences. *Science Education*	Inservice, elementary, USA (inferred)	Qual (I, O)	Epistemic affect
32	J	Dos Santos, F.M.T., & Mortimer, E.F.	2003	How emotions shape the relationship between a chemistry teacher and her high school students. *International Journal of Science Education*	Inservice, high school, Brazil	Qual (I, O)	Psych/Neuro
33	J	Dreon, O., & McDonald, S.	2012	Being in the hot spot: a phenomenological study of two beginning teachers' experiences enacting inquiry science pedagogy. *Teachers & Teaching: Theory and Practice*	Inservice, high school, USA	Qual (I, O, W)	Self-efficacy
34	J	Eilon, B., & Kliachko, S.	2004	Perceptions of the teacher's roles by prospective elementary school science teachers in a web-based biology course. *Journal of Technology & Teacher Education*	Preservice, elementary, Israel	Qual (I, W)	N/A
35	J	Ekborg, M.	2005	Is heat generated from a crematorium an appropriate source for district heating? Student teachers' reasoning about a complex environmental issue. *Environmental Education Research*	Inservice, elementary, Sweden	Qual (I)	N/A

(continued)

Table 2.2 (continued)

#	Pub type	Author	Year	Title/brief source details	Education context	Design (0ata)	Emotion theory
36	J	Erdemir, M., & Kandil İngeç, Ş.	2023	Correlation between energy metaphor and empathy tendency. *Journal of Baltic Science Education*	Preservice (mathematics, science), Turkey	Quant (SL)	Psych/linguistic
37	J	Fančovičová, J., Prokop, P., & Lešková, A.	2013	Perceived disgust and personal experiences are associated with acceptance of dissections in schools. *Eurasia Journal of Mathematics, Science and Technology Education*	Preservice, high school, Slovakia	Quant (SL)	Psychology
38	J	García-Ruiz, C., Lupión-Cobos, T., & Blanco-López, Á.	2022	Effects of an inquiry-based science education training program on pre-service teachers. A mixed-methods case study. *Eurasia Journal of Mathematics, Science and Technology Education*	Preservice, high school, Spain	Mixed methods (SL, W)	N/A
39	J	Gilbert, A., & Byers, C.C.	2017	Wonder as a tool to engage preservice elementary teachers in science learning and teaching. *Science Education*	Preservice, elementary, Australia	Qual (A, I, O, So)	N/A, (epistemic emotion)

(continued)

Table 2.2 (continued)

#	Pub type	Author	Year	Title/brief source details	Education context	Design (0ata)	Emotion theory
40	J	González-Weil, C., Reyes-Espejo, M.I.; Bravo González, P., Acuña Ruz, F., & Fernández Araneda, F.	2025	How do you trust in the midst of mistrust? A longitudinal study (2010–2021) on building trust in a professional learning community of science teachers in Chile. *Professional Development in Education*	Inservice, high school, Chile	Qual (FG, I)	N/A
41	J	Han-Tosunoglu, C., Özer-Aker, F., & Ağlarcı-Özdemir, O.	2025	Transformative shifts in informal reasoning of pre-service teachers in a real-time health issue. *Journal of Science Teacher Education*	Preservice, high school, turkey	Qual (I)	Education- SSI
42	J	Hawley, P.H., & Sinatra, G.M.	2019	Declawing the dinosaurs in the science classroom: Reducing Christian teachers' anxiety and increasing their efficacy for teaching evolution. *Journal of Research in Science Teaching*	Inservice, K-12, USA	Quant (SL)	Psychology, Self-efficacy
43	J	Hernández-Barco, M., Cañada-Cañada, F., Corbacho-Cuello, I., & Sánchez-Martín, J.	2021	An Exploratory Study Interrelating Emotion, Self-Efficacy and Multiple Intelligence of Prospective Science Teachers. *Frontiers*	Preservice, elementary, Spain	Quant (SL)	Psychology, Self-efficacy

(continued)

Table 2.2 (continued)

#	Pub type	Author	Year	Title/brief source details	Education context	Design (0ata)	Emotion theory
44	J	Hernandez- Barco M. A., Sanchez-Martín J., Corbacho-Cuello I., & Cañada-Cañada F.	2022	Is sustainability an exhausted concep?: Bridging the gap from environmental awareness to emotional proficiency in science education through integral sustainability. *Heliyon*	Preservice, elementary, Spain	Quant (SL)	Psychology, Self-efficacy
45	J	Hong, M., & Kang, N.H.	2010	South Korean and the US secondary school science teachers' conceptions of creativity and teaching for creativity. *International Journal of Science and Mathematics Education*	Inservice, middle/ high schools, South Korea, USA	Quant + qual (SL, So)	N/A
46	J	Hufnagel, E.	2015	Preservice elementary teachers' emotional connections and disconnections to climate change in a science course. *Journal of Research in Science Teaching*	Preservice, elementary, USA	Qual (A, O)	Discourse, Psychology
47	J	Jaber, L., Dini, V., & Hammer, D.	2021	"Well that's how the kids feel!"—Epistemic empathy as a driver of responsive teaching. *Journal of Research in Science Teaching*	Inservice, elementary, middle school USA	Qual (I, O)	N/A

(continued)

Table 2.2 (continued)

#	Pub type	Author	Year	Title/brief source details	Education context	Design (0ata)	Emotion theory
48	J	Jackson, D.F., Doster, E.C., Meadows, L., & Wood, T.	1995	Hearts and minds in the science classroom: The education of a confirmed evolutionist. *Journal of Research in Science Teaching*	Inservice, preservice, USA	Qual (I, FG, O)	N/A
49	J	Jeong, J.S., Cañada-Cañada, F., & González-Gómez, D.	2018	The study of flipped-classroom for pre-service science teachers.*Education Sciences*	Preservice, elementary, Spain	Quant (SL)	Constructionist (inferred)
50	J	Koul, R., Musikul, K., Nishikawa, S., & Almutawa, H.	2024	Mentoring in the development of science teaching self-efficacy among primary school teachers in Thailand: A Mixed Methods study. *Journal of Science Teacher Education*	Inservice, elementary, Thailand	Mixed methods (I, SL)	Self-efficacy
51	J	Kılınç, A., Kartal, T., Eroğlu, B.,Demiral, Ü., Afacan, Ö., Polat, D., Demirci G., Mutlu P., & Görgülü, Ö.	2013	Preservice science teachers' efficacy regarding a socioscientific issue: A belief system approach. *Research in Science Education*	Preservice, Turkey	Quant (SL)	Psychology
52	J	Kordíková, B., & Brestenská, B.	2022	Bilingual science education: perceptions of Slovak in-service and pre-service teachers. *International Journal of Bilingual Education and Bilingualism*	Inservice and Preservice, high school, Slovakia	Quant (SL)	N/A

(continued)

Table 2.2 (continued)

#	Pub type	Author	Year	Title/brief source details	Education context	Design (0ata)	Emotion theory
53	J	Lee, H., Chang, H., Choi, K., Kim, S.-W., & Zeidler, D.L.	2012	Developing character and values for global citizens: Analysis of pre-service science teachers' moral reasoning on socioscientific issues. *International Journal of Science Education*	Preservice, middle/high school (inferred), South Korea	Qual (FG, W)	Moral reasoning, ethic of care
54	J	Lombardi, D. & Sinatra, G.	2013	Emotions about teaching about human-induced climate change. *International Journal of Science Education*	Preservice secondary and elementary, USA	Quant + qual (I, O, SL)	Psychology/Education
55	J	López-Banet, L., Aguilera, D., Jiminéz-Liso, M. R., & Perales-Palacios, F. J.	2021	Emotional and cognitive preservice science teachers' engagement while living a model-based inquiry science technology engineering mathematics sequence about acid–base. *Frontiers*	Preservice, high school, Spain	Quant (SL)	Psychology/Education
56	J	Ma, H.	2018	Influence of the interplay of habitual affective attributes and classroom learning environments on learners' situational affective experiences in learning science: the narratives of primary pre-service teachers. *Research in Science Education*	Preservice, elementary, Australia	Qual (I)	Psychology

(continued)

Table 2.2 (continued)

#	Pub type	Author	Year	Title/brief source details	Education context	Design (0ata)	Emotion theory
57	J	Mahmood, M.S., Talafian, H., Shafer, D., Kuo, E., Lundsgaard, M., & Stelzer, T.	2024	Navigating socio-emotional risk through comfort-building in physics teacher professional development: A case study. *Journal of Research in Science Teaching*	Inservice, high school, USA	Qual (O)	Conceptual framework no single theory
58	J	Mbowane, C.K., de Villiers, J.J., Rian, B., & Max W.H.	2017	Teacher participation in science fairs as professional development in South Africa. *South African Journal of Science*	Inservice, high school, South Africa	Quant + qual (I, SL, So)	Identity
59	J	Melville, W, & Bartley, A.	2013	Constituting identities that challenge the contemporary discourse: power, discourse, experience, and emotion. *Science Education*	Inservice, high school, Canada	Qual (I)	Identity, post-structuralism
60	J	Melville, W., Bartley, A., & Fazio, X.	2013	Scaffolding the inquiry continuum and the constitution of identity. *International Journal of Science and Mathematics Education*	Preservice, middle/high school, Canada	Qual (I)	Identity, post-structuralism
61	J	Membiela, P., Acosta, K., & González, A.	2023	Antecedents of Chilean in-service science teachers' emotions: The important role of teacher motivation towards teaching career and self-efficacy. *European Journal of Teacher Education*	Inservice, Chile	Quant (SL)	Control Value Theory

(continued)

Table 2.2 (continued)

#	Pub type	Author	Year	Title/brief source details	Education context	Design (0ata)	Emotion theory
62	J	Mintzes, J.J., Marcum, B., Messerschmidt-Yates, C., & Mark, A.	2013	Enhancing self-efficacy in elementary science teaching with professional learning communities. *Journal of Science Teacher Education*	Inservice, elementary, USA	Mixed methods (E, I, SL)	Psychology
63	J	Moore Mensah, F.	2019	Finding voice and passion: Critical race theory methodology in science teacher education. *American Educational Research Journal*	Preservice, elementary, USA	Qual (A, I, O, W)	Critical race theory
64	J	Palmberg, I., Hermans, M., Jeronen, E., Kärkkäinen, S., Persson, C., & Yli-Panula, E.	2018	Nordic student teachers' views on the importance of species and species identification. *Journal of Science Teacher Education*	Preservice, elementary, Finland/Norway/ Sweden	Mixed methods (SL, So)	N/A
65	J	Polat, D., Uluay, G., & Başarmak, U.	2021	Examining the pre-service teachers' components of self-efficacy beliefs in science teaching (SEBST). *Sakarya University Journal of Education*	Preservice, not specified, Turkey	Quant (SL)	Self-efficacy
66	J	Praderio Gaias, F.N., Furman, M., Luzuriaga, M., Borrego, E.C., & Ayuso Fernández, G.E.	2024	The influence of emotions on science teaching: a case study with three early childhood education teachers. *Frontiers*	Inservice, early childhood, Argentina	Quant + qual (A, I, SL, So, O)	N/A
67	J	Rinchen, S., Ritchie, S.M., & Bellocchi, A.	2016	Emotional climate of a pre-service science teacher education class in Bhutan. *Cultural Studies of Science Education*	Preservice, middle/high school, Bhutan	Qual (SL)	Sociology

(continued)

Table 2.2 (continued)

#	Pub type	Author	Year	Title/brief source details	Education context	Design (0ata)	Emotion theory
68	J	Ritchie, S.M., Kidman, G., & Vaughan, T.	2007	Professional learning opportunities from uncovering cover stories of science and science teaching for a scientist-in-transition. *Cultural Studies of Science Education*	Preservice, high school, Australia	Qual (FG, W)	Identity
69	J	Ritchie, S.M., Tobin, K., Sandhu, M., Sandhu, S., Henderson, S., & Roth, W.-M.	2013	Emotional arousal of beginning physics teachers during extended experimental investigations. *Journal of Research in Science Teaching*	Preservice, high school, Australia	Qual (I, O)	Sociology
70	J	Ritchie, S.M., Tobin, K.G., Hudson, P., Roth, W.-M., & Mergard, V.	2011	Reproducing successful rituals in bad times: Exploring emotional interactions of a new science teacher. *Science Education*	Inservice, elementary, Australia	Qual (I, O, W)	Sociology
71	J	Rivera Maulucci, M.S.R.	2011	Language experience narratives and the role of autobiographical reasoning in becoming an urban science teacher. *Cultural Studies of Science Education*	Preservice, elementary, USA	Qual (A, I, W)	Multi-disciplinary-Sociology, Psych, Cognitive sci
72	J	Rivera Maulucci, M.S.R.	2013	Emotions and positional identity in becoming a social justice science teacher: Nicole's story. *Journal of Research in Science Teaching*	Pre-service, high school, USA	Qual (A, I, O)	Education

(continued)

Table 2.2 (continued)

#	Pub type	Author	Year	Title/brief source details	Education context	Design (0ata)	Emotion theory
73	J	Rydin, M.; Gericke, N.; Christenson, N.; Haglund, J.	2025	Areas of tension in teaching climate change–conflicts between Swedish preservice science teachers' beliefs. *International Journal of Science Education*	Preservice, high school, Sweden	Qual (I)	N/A
74	J	Sanchez-Martin, J., Cañada-Cañada, F., & Dávila-Acedo, A.	2018	Emotional responses to innovative science teaching methods: Acquiring emotional data in a general science teacher education Class. *Journal of Technology and Science Education*	Preservice, elementary, Spain	Quant (SL)	Education
75	J	Sullenger, K., Turner, S., Caplan, H., Crummey, J., Cuming, R., Charron, C., & Corey, B.	2000	Culture wars in the classroom: prospective teachers question science. *Journal of Research in Science Teaching*	Preservice, elementary, USA	Qual (W)	N/A
76	J	Tang, X., Elby, A., Levin, D., & Chumbley, A. K.	2025	Complex dynamics behind the scene: How identity and affect shape epistemic stance in science teaching. *Cognition and Instruction*	Preservice/ Inservice, middle school, USA	Qual (I, O, W)	Affect, Psych
77	J	Thomson, M.M., & Turner, J.E.	2019	The role of emotions in teachers' professional development: Attending a research experience for teachers (RET) Program. *Education Research International*	Inservice, elementary, middle, high school, USA	Quant (SL)	Psychology

(continued)

Table 2.2 (continued)

#	Pub type	Author	Year	Title/brief source details	Education context	Design (0ata)	Emotion theory
78	J	Tobin, K., Ritchie, S. M., Oakley, J., Mergard, V., & Hudson, P. B.	2013	Relationship between emotional climate and the fluency of classroom interactions. *Learning Environments Research*	Inservice, elementary, Australia	Qual (O)	Sociology, Psych
79	J	Topçu, M.S., Yilmaz-Tuzun, O., & Sadler, T.D.	2011	Turkish preservice science teachers' informal reasoning regarding socioscientific issues and the factors influencing their informal reasoning. *Journal of Science Teacher Education*	Preservice, middle school, Turkey	Quant (SL)	Informal reasoning, Psychology
80	J	Tuncay, B., Yılmaz-Tüzün, Ö., & Teksoz, G.-T.	2012	Moral reasoning patterns and influential factors in the context of environmental problems. *Environmental Education Research*	Preservice, Turkey	Mixed methods (I, SL)	Moral reasoning
81	J	Uzuntiryaki-Kondakci, E., Kirbulut, Z.D., Oktay, O., & Sarici, E.	2022	A qualitative examination of science teachers' emotions, emotion regulation goals and strategies. *Research in Science Education*	Inservice, middle schools, Middle Turkey	Qual (I, O, W)	Control Value theory
82	J	van Rooij, E., Cornelia M., Fokkens-Bruinsma, M., & Goedhart, M.	2019	Preparing Science undergraduates for a teaching career: Sources of their teacher self-efficacy. *Teacher Educator*	Preservice science minor, Dutch students (Netherlands)	Quant (SL)	Psychology, Self-efficacy

(continued)

Table 2.2 (continued)

#	Pub type	Author	Year	Title/brief source details	Education context	Design (0ata)	Emotion theory
83	J	Velasco, R. C. L., Hite, R., & Milbourne, J.	2022	Exploring advocacy self-efficacy among k-12 stem teacher leaders. *International Journal of Science and Mathematics Education*	Inservice, elementary, secondary, USA	Qual (A, I, FG)	Self-efficacy
84	J	Wagler, R., & Wagler, A.	2017	Understanding how preservice teachers' fear, perceived danger and disgust affects the incorporation of arachnid information into the elementary science classroom. *International Journal of Environmental and Science Education*	Preservice, middle school, USA	Quant (SL)	Behavior/Evolutionary
85	J	Wagler, R., & Wagler, A.	2022	A randomized and controlled research study assessing the emotions and beliefs of future middle school science teachers toward terrestrial isopods.*Insects*	Preservice, middle school, USA (inferred)	Quant (E, SI)	Psychology, Biology
86	J	Wicaksono, A.G.C., & Korom, E.	2023	Attitudes towards science in higher education: Validation of questionnaire among science teacher candidates and engineering students in Indonesia. *Heliyon*	Preservice (science and eng.), Central Java, Indonesia	Quant (SL)	Communication

(continued)

Table 2.2 (continued)

#	Pub type	Author	Year	Title/brief source details	Education context	Design (0ata)	Emotion theory
87	J	Yerdelen, S., Osmanoglu, A., & Tas, Y.	2019	The influence of a teaching practice course with video-case enriched microteaching on prospective teachers' self-efficacy for teaching. *International Journal of Research in Education and Science*	Preservice, high school, Turkey	Quant (SL)	Psychology, Self-efficacy
88	J	YüCel, E. Ö, & Özkan, M.	2018	Analysis of change in the environmental perceptions of prospective science teachers: A longitudinal study. *Asia–Pacific Forum on Science Learning and Teaching*	Preservice, high school (inferred), Turkey	Q uant (SL)	Psychology, Behaviour
89	J	Zembylas, M.	2004a	Emotion metaphors and emotional labor in science teaching. *Science Education*	Inservice, elementary, USA	Qual (A, I, O, W),	Multidisciplinary-Sociology, Psychology, & Cognitive sci
90	J	Zembylas, M.	2002	Constructing genealogies of teachers' emotions in science teaching. *Journal of Research in Science Teaching*	Inservice, elementary, USA	Qual (A, I, O, W)	Multidisciplinary-Sociology, Psychology, & Cognitive sci

(continued)

Table 2.2 (continued)

#	Pub type	Author	Year	Title/brief source details	Education context	Design (Data)	Emotion theory
91	J	Zembylas, M., & Buhlman Barker, H.	2002	Preservice teacher attitudes and emotions: Individual spaces, community conversations and transformations. *Research in Science Education*	Preservice, elementary, USA	Qual (A, W, I, O)	Education/Cultural/ Feminist perspectives
92	J	Zembylas, M	2004b	Emotional issues in teaching science: A case study of a teacher's views. *Research in Science Education*	Inservice, elementary, USA	Qual (A, I, O)	Social constructionist

Key:
Publication Types: B = Book chapter; J = Journal article
Methods: A = artifacts (e.g., objects, lesson plans), E = experimental/randomized control trial, I = interviews, FG = focus group interviews, O = observations, So = open-ended survey, SL = Likert survey, It = introspection

author conducted a narrative synthesis of the extracted data, organizing studies thematically based on their primary focus (e.g., controversial issues, teacher self-efficacy and identity; see Chaps. 4–9). This approach was chosen due to the heterogeneity of study designs, methodologies, and outcomes that precluded a meta-analysis.

In the following chapters, we present the findings of our review, beginning in Chap. 3 with a general overview, including research contexts, samples, and study designs of empirical studies of emotion in science teacher education. We also construct six themes developed through our analysis which serve to group included studies according to their foci and subsequently organize Chaps. 4–9.

References

NB Articles included in the systematic review are indicated using an asterisk (*)

*Abualrob, M. M. A., & Al-Saadi, S. H. (2025). Evaluating the role of technological pedagogical content knowledge in ICT-enhanced science education. *International Journal of Information and Education Technology, 15*, 117–126. https://doi.org/10.18178/ijiet.2025.15.1.2224

*Adler, I., Zion, M., & Rimerman-Shmueli, E. (2019). Fostering teachers' reflections on the dynamic characteristics of open inquiry through metacognitive prompts. *Journal of Science Teacher Education, 30*, 763–787. https://doi.org/10.1080/1046560X.2019.1627060

*Akerson, V. L., Pongsanon, K., Weiland, I. S., & Nargund-Joshi, V. (2014). Developing a professional identity as an elementary teacher of nature of science: A self-study of becoming an elementary teacher. *International Journal of Science Education, 36*, 2055–2082. https://doi.org/10.1080/09500693.2014.890763

*Alarcón-Orozco, M. M., Franco-Mariscal, A. J., Oliva, J. M., & Blanco-López, Á. (2025). Emotions experienced by preservice early childhood teachers during a training program in inquiry-based science education. *Journal of Research in Science Teaching*. https://doi.org/10.1002/tea.70007

*Amponsah, K. D., Adu-Gyamfi, K., Awoniyi, F.C., & Commey-Mintah, P. (2024). Navigating academic performance: Unravelling the relationship between emotional intelligence, learning styles, and science and technology self-efficacy among preservice science teachers. *Heliyon, 10*(9). https://doi.org/10.1016/j.heliyon.2024.e29474

*Avraamidou, L. (2018). Elementary science teacher identity as a lived experience: small stories in narrative analysis. In P. A. Schutz et al. (eds.), *Research on teacher identity* (pp. 145–155). Springer. https://doi.org/10.1007/978-3-319-93836-3_13

*Bahcivan, E. (2019). Examining the structural relations among PSTs' scientific epistemological beliefs, epistemic emotions and argumentativeness: Sample from Turkey. *International Journal of Education in Mathematics, Science and Technology, 7*(3), 271–280.

*Bellocchi, A., Davis, J., & King, D. T. (2018a). Science demonstrations as mediators of interaction ritual chains: Temporal transitions from mundane to intense emotional energy. In S. M. Ritchie & K. G. Tobin (Eds.), *Eventful learning: Learner emotions* (pp. 57–86). Brill. https://doi.org/10.1163/9789004377912_004

*Bellocchi, A., Mills, K. A., & Ritchie, S. M. (2016). Emotional experiences of preservice science teachers in online learning: The formation, disruption and maintenance of social bonds. *Cultural Studies of Science Education, 11*, 629–652. https://doi.org/10.1007/s11422-015-9673-9

*Bellocchi, A., Mills, K., Olson, R., Patulny, R., & McKenzie, J. (2018b). Emotion work at the frontline of STEM teaching. In L. Bryan & K. Tobin (Eds.), *Critical issues and bold visions for science education: The road ahead* (pp. 247–264). Brill. https://doi.org/10.1163/9789004389663_014

*Bellocchi, A., Ritchie, S. M., Tobin, K., Sandhu, M., & Sandhu, S. (2013). Exploring emotional climate in pre-service science teacher education. *Cultural Studies in Science Education, 8*, 529–552. https://doi.org/10.1007/s11422-013-9526-3

*Bellocchi, A., Ritchie, S. M., Tobin, K., King, D., Sandhu, M., & Henderson, S. (2014). Emotional climate and high quality learning experiences in science teacher education. *Journal of Research in Science Teaching, 51*, 1301–1325. https://doi.org/10.1002/tea.21170

*Cebesoy, Ü. B. (2025). Exploring preservice science teachers' emotive reasoning through place-based SSI instruction: The case of gold mining. In B. Namdar & E. Karahan (Eds.), *Socioscientific issues focused teacher education* (pp.153–180). Springer. https://doi.org/10.1007/978-3-031-55233-5_10

*Cebesoy, Ü. B., & Chang Rundgren, S. N. (2023). Embracing socioscientific issues-based teaching and decision-making in teacher professional development. *Educational Review, 75*(3), 507–534. https://doi.org/10.1080/00131911.2021.1931037

*Borrachero, A. B., Brígido, M., Mellado, L., Costillo, E., & Mellado, V. (2014). Emotions in prospective secondary teachers when teaching science content, distinguishing by gender. *Research in Science and Technology Education, 32*(2). https://doi.org/10.1080/02635143.2014.909800

*Brick, K., Cooper, J., Mason, L., Faeflen, S., Monmia, J., & Dubinsky, J. M. (2021). Tiered neuroscience and mental health professional development in Liberia improves teacher self-efficacy, self-responsibility, and motivation. *Frontiers in Human Neuroscience, 15*, 664730. https://doi.org/10.3389/fnhum.2021.664730

*Brígido, M., Bermejo, M. L., Conde, M. C., & Mellado, V. (2010). The emotions in teaching and learning nature sciences and physics/chemistry in pre-service primary teachers. *US-China Educational Review, 7*(12), 25–32.

*Brígido, M., Borrachero, A., Bermejo, M. L., & Mellado, V. (2013a). Prospective primary teachers' self-efficacy and emotions in science teaching. *European Journal of Teacher Education, 36*, 1–18. https://doi.org/10.1080/02619768.2012.686993

*Brígido, M., Couso, D., Gutiérrez, C., & Mellado, V. (2013b). The emotions about teaching and learning science: A study of prospective primary teachers in three Spanish universities. *Journal of Baltic Science Education, 12*(3), 299–311.

*Bullough, R.V. (2009). Seeking eudaimonia: The emotions in learning to teach and to mentor. In Schutz, A., & Zembylas, M. (Eds.), *Advances in teacher emotion research* (pp.33–54). Springer. DI: https://doi.org/10.1007/978-1-4419-0564-2_3

*Can, H. (2015). Sources of teaching efficacy beliefs in pre-service science teachers. *Elementary Education Online, 14*(3), 1044–1059.

*Cansoy, R. (2017). Teachers' professional development: The case of WhatsApp. *Journal of Education and Learning, 6*(4), 285–293. https://doi.org/10.5539/jel.v6n4p285

*Cook, K., & Buck, G. (2017). Our neighbourhood: A place for heightening emotional energy in science education. In A. Bellocchi, C. Quigley, & K. Otrel-Cass (Eds.), *Exploring emotions, aesthetics and wellbeing in science education research* (pp. 141–164). Springer. https://doi.org/10.1007/978-3-319-43353-0_8

*Cooke, H., Campbell, T., Luehmann, A., Zhang, Y., & Scipio, D. (2025). "I sit here thinking I can do this"—Developing justice-centered ambitious science teaching identities in professional learning communities. *Journal of Research in Science Teaching, 64*(2), 1040–1072. https://doi.org/10.1002/tea.21991

*Cross, D. I., & Hong, J.-Y. (2009). Beliefs and professional identity: Critical constructs in examining the impact of reform on the emotional experiences of teachers. In A. Schutz & M. Zembylas (Eds.), *Advances in teacher emotion research.* Springer. https://doi.org/10.1007/978-1-4419-0564-2_14

*Davidson, S. G., Jaber, L. Z., & Southerland, S. A. (2020). Emotions in the doing of science: Exploring epistemic affect in elementary teachers' science research experiences. *Science Education, 104*, 1008–1040. https://doi.org/10.1002/sce.21596

*Dos Santos, F. M. T., & Mortimer, E. F. (2003). How emotions shape the relationship between a chemistry teacher and her high school students. *International Journal of Science Education, 25*(9), 1095–1110. https://doi.org/10.1080/0950069032000052216

*Dreon, O., & McDonald, S. (2012). Being in the hot spot: A phenomenological study of two beginning teachers' experiences enacting inquiry science pedagogy. *Teachers & Teaching: Theory and Practice, 18*(3), 297–313.

*Eilon, B., & Kliachko, S. (2004). Perceptions of the teacher's roles by prospective elementary school science teachers in a web-based biology course. *Journal of Technology & Teacher Education, 12*(3), 339–360.

*Ekborg, M. (2005). Is heat generated from a crematorium an appropriate source for district heating? Student teachers' reasoning about a complex environmental issue. *Environmental Education Research, 11*(5), 557–573. https://doi.org/10.1080/13504620500169718

*Erdemir, M., & Kandil İngeç, Ş. (2023). Correlation between energy metaphor and empathy tendency. *Journal of Baltic Science Education, 22*(5), 813–832. https://doi.org/10.33225/jbse/23.22.813

*Fančovičová, J., Prokop, P., & Lešková, A. (2013). Perceived disgust and personal experiences are associated with acceptance of dissections in schools. *Eurasia Journal of Mathematics, Science and Technology Education, 9*(3), 311–318. https://doi.org/10.12973/eurasia.2013.938a

*García-Ruiz, C., Lupión-Cobos, T., & Blanco-López, Á. (2022). Effects of an inquiry-based science education training program on pre-service teachers. A mixed-methods case study. *Eurasia Journal of Mathematics, Science and Technology Education, 18*(12), em2186. https://doi.org/10.29333/ejmste/12578

*Gilbert, A., & Byers, C. C. (2017). Wonder as a tool to engage preservice elementary teachers in science learning and teaching. *Science Education, 101*, 907–928. https://doi.org/10.1002/sce.21300

*González-Weil, C., Reyes-Espejo, M. I., Bravo González, P., Acuña Ruz, F., & Fernández Araneda, F. (2025). How do you trust in the midst of mistrust? A longitudinal study (2010–2021) on building trust in a professional learning community of science teachers in Chile. *Professional Development in Education, 51*(1), 39–53. https://doi.org/10.1080/19415257.2024.2431701

Grant, M. J., & Booth, A. (2009). A typology of reviews: An analysis of 14 review types and associated methodologies. *Health Information and Libraries Journal, 26*, 91–108. https://doi.org/10.1111/j.1471-1842.2009.00848.x

*Han-Tosunoglu, C., Özer-Aker, F., & Ağlarcı-Özdemir, O. (2025). Transformative shifts in informal reasoning of pre-service teachers in a real-time health issue. *Journal of Science Teacher Education, 36*(4), 1–35. https://doi.org/10.1080/1046560X.2024.2404339

*Hawley, P. H., & Sinatra, G. M. (2019). Declawing the dinosaurs in the science classroom: Reducing Christian teachers' anxiety and increasing their efficacy for teaching evolution. *Journal of Research in Science Teaching, 56*(4), 375–401. https://doi.org/10.1002/tea.21479

*Hernandez-Barco, M.A., Sanchez-Martín J., Corbacho-Cuello I., & Cañada-Cañada F. (2022). Is sustainability an exhausted concept: Bridging the gap from environmental awareness to emotional proficiency in science education through integral sustainability. *Heliyon, 8*(12). https://doi.org/10.1016/j.heliyon.2022.e12546

*Hernández-Barco, M., Cañada-Cañada, F., Corbacho-Cuello, I., & Sánchez-Martín, J. (2021). An exploratory study interrelating emotion, self-efficacy and multiple intelligence of prospective science teachers. *Frontiers in Education, 6*, Article 604791. https://doi.org/10.3389/feduc.2021.604791

*Hobbs, L., & Kelly, L. (2017). The heart of the educator: Aesthetic experience shaping knowledge, identity, and passion. In A. Bellocchi, C. Quigley, & K. Otrel-Cass (Eds.), *Exploring emotions, aesthetics and wellbeing in science education research*. Springer. https://doi.org/10.1007/978-3-319-43353-0_4

*Hong, M., & Kang, N. H. (2010). South Korean and the US secondary school science teachers' conceptions of creativity and teaching for creativity. *International Journal of Science and Mathematics Education, 8*, 821–843. https://doi.org/10.1007/s10763-009-9188-5

*Hufnagel, E. (2015). Preservice elementary teachers' emotional connections and disconnections to climate change in a science course. *Journal of Research in Science Teaching, 52*, 1296–1324. https://doi.org/10.1002/tea.21245
*Jaber, L., Dini, V., & Hammer, D. (2021). "Well that's how the kids feel!"—Epistemic empathy as a driver of responsive teaching. *Journal of Research in Science Teaching, 59*, 223–251. https://doi.org/10.1002/tea.21726
*Jackson, D. F., Doster, E. C., Meadows, L., & Wood, T. (1995). Hearts and minds in the science classroom: The education of a confirmed evolutionist. *Journal of Research in Science Teaching, 32*, 585–611. https://doi.org/10.1002/tea.3660320606
*Jeong, J. S., Cañada-Cañada, F., & González-Gómez, D. (2018). The study of flipped-classroom for pre-service science teachers. *Education Sciences, 8*(4), 1630. https://doi.org/10.3390/educsci8040163
*Kılınç, A., Kartal, T., Eroğlu, B., Demiral, Ü., Afacan, Ö., Polat, D., Demirci, G., Mutlu, P., & Görgülü, Ö. (2013). Preservice science teachers' efficacy regarding a socioscientific issue: A belief system approach. *Research in Science Education, 43*, 2455–2475. https://doi.org/10.1007/s11165-013-9368-8
*Kordíková, B., & Brestenská, B. (2022). Bilingual science education: Perceptions of Slovak in-service and pre-service teachers. *International Journal of Bilingual Education and Bilingualism, 25*(3), 1–14. https://doi.org/10.1080/13670050.2020.1718590
*Koul, R., Musikul, K., Nishikawa, S., & Almutawa, H. (2024). Mentoring in the development of science teaching self-efficacy among primary school teachers in Thailand: A Mixed Methods study. *Journal of Science Teacher Education, 35*(1), 44–62. https://doi.org/10.1080/1046560X.2023.2179384
*Lee, H., Chang, H., Choi, K., Kim, S.-W., & Zeidler, D. L. (2012). Developing character and values for global citizens: Analysis of pre-service science teachers' moral reasoning on socioscientific issues. *International Journal of Science Education, 34*(6), 925–953. https://doi.org/10.1080/09500693.2011.625505
*Lombardi, D., & Sinatra, G. (2013). Emotions about teaching about human-induced climate change. *International Journal of Science Education, 35*(1), 167–191. https://doi.org/10.1080/09500693.2012.738372
*López-Banet, L., Aguilera, D., Jiminéz-Liso, M. R., & Perales-Palacios, F. J. (2021). Emotional and cognitive preservice science teachers' engagement while living a model-based inquiry science technology engineering mathematics sequence about acid-base. *Frontiers in Psychology, 12*, Article 719648. https://doi.org/10.3389/fpsyg.2021.719648
*Ma, H. (2018). Influence of the interplay of habitual affective attributes and classroom learning environments on learners' situational affective experiences in learning science: The narratives of primary pre-service teachers. *Research in Science Education, 51*, 399–417. https://doi.org/10.1007/s11165-018-9800-1
*Mahmood, M. S., Talafian, H., Shafer, D., Kuo, E., Lundsgaard, M., & Stelzer, T. (2024). Navigating socio-emotional risk through comfort-building in physics teacher professional development: A case study. *Journal of Research in Science Teaching, 61*, 2223–2257. https://doi.org/10.1002/tea.21949
*Mbowane, C. K., de Villiers, J. J., Rian, B., & Max, W. H. (2017). Teacher participation in science fairs as professional development in South Africa. *South African Journal of Science, 113*(7/8), 7. https://doi.org/10.17159/sajs.2017/20160364
*Melville, W., & Bartley, A. (2013). Constituting identities that challenge the contemporary discourse: Power, discourse, experience, and emotion. *Science Education, 97*, 171–190. https://doi.org/10.1002/sce.21047
*Melville, W., Bartley, A., & Fazio, X. (2013). Scaffolding the inquiry continuum and the constitution of identity. *International Journal of Science and Mathematics Education, 11*, 1255–1273. https://doi.org/10.1007/s10763-012-9375-7
*Membiela, P., Acosta, K., & González, A. (2023). Antecedents of Chilean in-service science teachers' emotions: The important role of teacher motivation towards teaching career and

self-efficacy. *European Journal of Teacher Education, 48*(2), 1–20. https://doi.org/10.1080/02619768.2023.2229003
*Mintzes, J. J., Marcum, B., Messerschmidt-Yates, C., & Mark, A. (2013). Enhancing self-efficacy in elementary science teaching with professional learning communities. *Journal of Science Teacher Education, 24*, 1201–1218. https://doi.org/10.1007/s10972-012-9320-1
*Moore Mensah, F. (2019). Finding voice and passion: Critical race theory methodology in science teacher education. *American Educational Research Journal, 65*(4). https://doi.org/10.3102/0002831218818093.
Page, M. J., McKenzie J. E., Bossuyt P. M., Boutron, I., Hoffmann T. C., Mulrow, C. D., et al. (2021). The PRISMA 2020 statement: an updated guideline for reporting systematic reviews. *British Medical Journal, 372*(71). https://doi.org/10.1136/bmj.n71
*Palmberg, I., Hermans, M., Jeronen, E., Kärkkäinen, S., Persson, C., & Yli-Panula, E. (2018). Nordic student teachers' views on the importance of species and species identification. *Journal of Science Teacher Education, 29*, 397–419. https://doi.org/10.1080/1046560X.2018.1468167
*Polat, D., Uluay, G., & Başarmak, U. (2021). Examining the pre-service teachers' components of self-efficacy beliefs in science teaching (SEBST). *Sakarya University Journal of Education, 11*(1), 195–217. https://doi.org/10.19126/suje.798077
*Praderio Gaias, F. N., Furman, M., Luzuriaga, M., Borrego, E. C., & Ayuso Fernández, G. E. (2024). The influence of emotions on science teaching: A case study with three early childhood education teachers. *Frontiers in Education, 9*, 1472721. https://doi.org/10.3389/feduc.2024.1472721
*Rinchen, S., Ritchie, S. M., & Bellocchi, A. (2016). Emotional climate of a pre-service science teacher education class in Bhutan. *Cultural Studies of Science Education, 11*, 1–26. https://doi.org/10.1007/s11422-014-9658-0
*Ritchie, S. M., & Newlands, J. (2017). Emotional events in learning science. In A. Bellocchi, C. Quigley, & K. Otrel-Cass (Eds.), *Exploring emotions, aesthetics and wellbeing in science education research* (pp. 107–121). Springer. https://doi.org/10.1007/978-3-319-43353-0_6
*Ritchie, S. M., Kidman, G., & Vaughan, T. (2007). Professional learning opportunities from uncovering cover stories of science and science teaching for a scientist-in-transition. *Cultural Studies of Science Education, 2*, 225–242. https://doi.org/10.1007/s11422-006-9044-7
*Ritchie, S. M., Tobin, K., Sandhu, M., Sandhu, S., Henderson, S., & Roth, W.-M. (2013). Emotional arousal of beginning physics teachers during extended experimental investigations. *Journal of Research in Science Teaching, 50*, 137–161. https://doi.org/10.1002/tea.21060
*Ritchie, S. M., Tobin, K. G., Hudson, P., Roth, W.-M., & Mergard, V. (2011). Reproducing successful rituals in bad times: Exploring emotional interactions of a new science teacher. *Science Education, 95*, 745–765. https://doi.org/10.1002/sce.20440
*Rivera Maulucci, M. S. (2012). Exploring linkages between identity and emotions in teaching for social justice in science teacher education. In M. Varelas (Ed.), *Identity construction and science education research: Learning, teaching, and being in multiple contexts* (pp. 123–139). Sense Publishers. https://doi.org/10.1007/978-94-6209-043-9_9
*Rivera Maulucci, M. S. R. (2011). Language experience narratives and the role of autobiographical reasoning in becoming an urban science teacher. *Cultural Studies of Science Education, 6*, 413–434. https://doi.org/10.1007/s11422-011-9323-9
*Rivera Maulucci, M. S. R. (2013). Emotions and positional identity in becoming a social justice science teacher: Nicole's story. *Journal of Research in Science Teaching, 50*, 453–478. https://doi.org/10.1002/tea.21081
Rosiek, J., & Beghetto, R. A. (2009). Emotional scaffolding: The emotional and imaginative dimensions of teaching and learning. In P. Schutz & M. Zembylas (Eds.), *Advances in teacher emotion research: The impacts on teachers' lives* (pp. 175–194). Springer.
*Rydin, M., Gericke, N., Christenson, N., & Haglund, J. (2025). Areas of tension in teaching climate change–conflicts between Swedish preservice science teachers' beliefs. *International Journal of Science Education*, 1–25. https://doi.org/10.1080/09500693.2025.2466771

*Sanchez-Martin, J., Cañada-Cañada, F., & Dávila-Acedo, A. (2018). Emotional responses to innovative science teaching methods: Acquiring emotional data in a general science teacher education Class. *Journal of Technology and Science Education, 8*(4), 346. https://doi.org/10.3926/jotse.408

*Şevki Ayvaci, H. Ş, Bebek, G., & Yamaçli, S. (2023). Experiences of pre-service science teachers in "teaching practice" during the sCovid-19 pandemic. *International e-Journal of Educational Studies, 13*, 134–152. https://doi.org/10.31458/iejes.1229871

Shapiro, B. (2005). From despair to success: A case study of support and transformation in an elementary science practicum. In S. Alsop (Ed.), *Beyond Cartesian dualism: Encountering affect in the teaching and learning of science* (pp. 161–172). Springer.

Sinatra, G. M., Broughton, S. H., & Lombardi, D. (2014). Emotions in science education. In R. Pekrun & L. Linnenbrink-Garcia (Eds.), *International handbook of emotions in education* (pp. 415–436). Routledge.

*Sullenger, K., Turner, S., Caplan, H., Crummey, J., Cuming, R., Charron, C., & Corey, B. (2000). Culture wars in the classroom: Prospective teachers question science. *Journal of Research in Science Teaching, 37*, 895–915. https://doi.org/10.1002/1098-2736(200011)37:9%3c895::AID-TEA3%3e3.0.CO;2-3

*Tang, X., Elby, A., Levin, D. M., & Chumbley, A. K. (2025). Complex dynamics behind the scene: How identity and affect shape epistemic stance in science teaching. *Cognition and Instruction, 43*, 134–173. https://doi.org/10.1080/07370008.2025.2473070

Thoits, P. A. (1989). The sociology of emotions. *Annual Review of Sociology, 15*, 317–342.

*Thomson, M. M., & Turner, J. E. (2019). The role of emotions in teachers' professional development: Attending a research experience for teachers (RET) Program. *Education Research International*. https://doi.org/10.1155/2019/5856793

*Tobin, K., Ritchie, S. M., Oakley, J., Mergard, V., & Hudson, P. B. (2013). Relationship between emotional climate and the fluency of classroom interactions. *Learning Environments Research, 16*, 71–89. https://doi.org/10.1007/s10984-013-9125-y

*Topçu, M. S., Yilmaz-Tuzun, O., & Sadler, T. D. (2011). Turkish preservice science teachers' informal reasoning regarding socioscientific issues and the factors influencing their informal reasoning. *Journal of Science Teacher Education, 22*, 313–332. https://doi.org/10.1007/s10972-010-9221-0

*Tuncay, B., Yılmaz-Tüzün, Ö., & Teksoz, G.-T. (2012). Moral reasoning patterns and influential factors in the context of environmental problems. *Environmental Education Research, 18*(4), 485–505. https://doi.org/10.1080/13504622.2011.630576

*Uzuntiryaki-Kondakci, E., Kirbulut, Z. D., Oktay, O., & Sarici, E. (2022). A qualitative examination of science teachers' emotions, emotion regulation goals and strategies. *Research in Science Education, 52*, 1131–1155. https://doi.org/10.1007/s11165-020-09986-y

*van Rooij, E., Cornelia, M., Fokkens-Bruinsma, M., & Goedhart, M. (2019). Preparing Science undergraduates for a teaching career: Sources of their teacher self-efficacy. *Teacher Educator, 54*(3), 270–294. https://doi.org/10.1080/08878730.2019.1606374

*Velasco, R. C. L., Hite, R., & Milbourne, J. (2022). Exploring advocacy self-efficacy among k-12 stem teacher leaders. *International Journal of Science and Mathematics Education, 20*, 435–457. https://doi.org/10.1007/s10763-021-10176-z

*Wagler, R., & Wagler, A. (2017). Understanding how preservice teachers' fear, perceived danger and disgust affects the incorporation of arachnid information into the elementary science classroom. *International Journal of Environmental and Science Education, 12*(2), 213–231.

*Wagler, R., & Wagler, A. (2022). A randomized and controlled research study assessing the emotions and beliefs of future middle school science teachers toward terrestrial isopods. *Insects, 13*(3), 233. https://doi.org/10.3390/insects13030233

*Wicaksono, A. G. C., & Korom, E. (2023). Attitudes towards science in higher education: Validation of questionnaire among science teacher candidates and engineering students in Indonesia. *Heliyon, 9*(9), Article e20023. https://doi.org/10.1016/j.heliyon.2023.e20023

*Yerdelen, S., Osmanoglu, A., & Tas, Y. (2019). The influence of a teaching practice course with video-case enriched microteaching on prospective teachers' self-efficacy for teaching. *International Journal of Research in Education and Science, 5*(2), 560–573.

*YüCel, E. Ö., & Özkan, M. (2018). Analysis of change in the environmental perceptions of prospective science teachers: A longitudinal study. *Asia-Pacific Forum on Science Learning and Teaching, 19*(1).

*Zembylas, M. (2001). A paralogical affirmation of emotion's discourse in science teaching. In A. Barton & M. Osborne (Eds.), *Teaching science in diverse settings: Marginalized discourses and classroom practice* (pp. 99–128). Peter Lang.

*Zembylas, M. (2002). Constructing genealogies of teachers' emotions in science teaching. *Journal of Research in Science Teaching, 39*, 79–103. https://doi.org/10.1002/tea.10010

*Zembylas, M. (2004a). Emotion metaphors and emotional labor in science teaching. *Science Education, 88*, 301–324. https://doi.org/10.1002/sce.10116

*Zembylas, M. (2004b). Emotional issues in teaching science: A case study of a teacher's views. *Research in Science Education, 34*, 343–364. https://doi.org/10.1007/s11165-004-0287-6

Zembylas, M. (2005a). Emotions and science teaching: Present research and future agendas. In S. Alsop (Ed.), *Beyond Cartesian dualism: Encountering affect in the teaching and learning of science* (pp. 123–132). Springer.

*Zembylas, M., & Buhlman Barker, H. (2002). Preservice teacher attitudes and emotions: Individual spaces, community conversations and transformations. *Research in Science Education, 32*, 329–351. https://doi.org/10.1023/A:1020862000107

Chapter 3
General Publication Trends on Emotion in Science Teacher Education

The first recorded publication in our database is published in 1995, with the next article appearing 5 years later. As shown in Fig. 3.1, a steady increase in articles published per annum between 2000 and 2007 takes place, until an exponential increase occurs which peaks in 2013 and again in 2019 after an interim slump. From 2016 another sharp rise occurred in the three years to 2019. Over the 30-year period shown in Fig. 3.1, the figures represent an average publication rate of 4 articles per annum. Focusing on the decade between 2015 and 2024, the publication rate increases to 5 per annum, compared to the previous decade 2013–2014, where the average is 3.2 per annum. These patterns show a growing and sustained interest in the topic of science teacher emotions, reflecting a vibrant and sustained field of inquiry.

The overall analysis of article characteristics reveals interesting patterns in the focus of emotions research in preservice and inservice science teacher education as summarized in Table 3.1. Considering the searches are limited to English language articles, it is unsurprising to find the largest number of studies conducted in the United States of America (USA), which is consistent with the origin of most science education research, more broadly as seen in the systematic reviews discussed in Chap. 1 (cf. Lin et al., 2025; Tosun, 2024) Notwithstanding the limitation to English language articles, Table 3.1 shows that studies have been conducted globally across Western and Eastern contexts. Despite the wide geographical distribution of studies, it is evident from an analysis of key theories informing included studies (discussed later) are based in Western theoretical traditions, bringing into question the validity of seeking to understand the emotional experiences of preservice and inservice teachers in countries founded on different philosophical foundations (e.g., Bhutan, Saudi Arabia, Turkey, South Korea). This issue of the potential mismatch between Western theory of emotion and Eastern preservice teachers' learning is noted by Rinchen et al. (2016) who applied Western emotion theories to understand the learning experiences of preservice physics teachers in relation to class emotional climate in Bhutan. Although the theories were useful for interpreting the data, Rinchen et al. (2016)

A. Bellocchi et al., *Emotion Research in Science Teacher Education*, Sociocultural Explorations of Science Education 32, https://doi.org/10.1007/978-3-032-07711-0_3

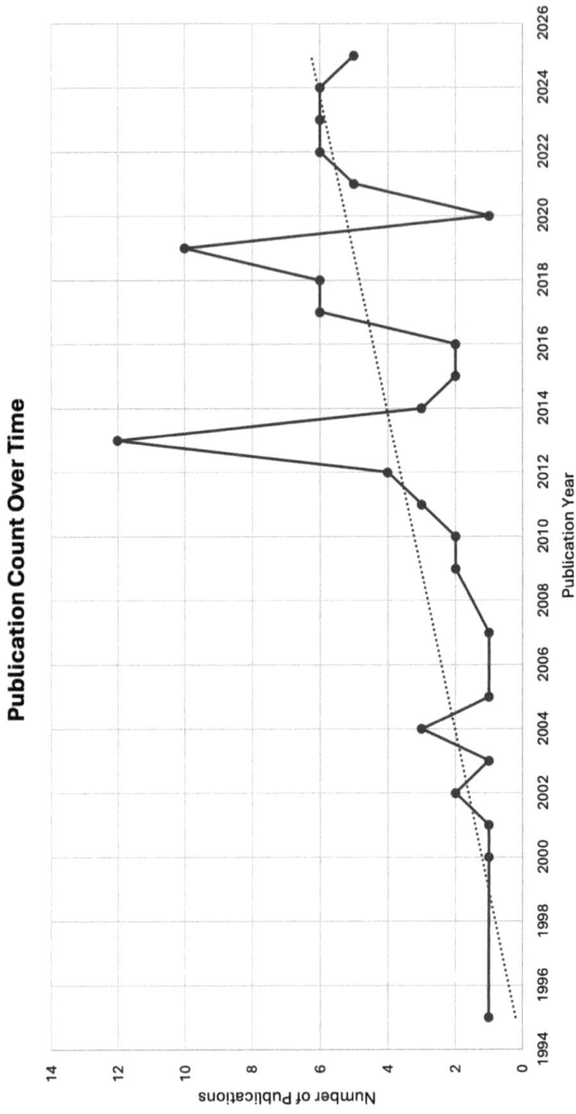

Fig. 3.1 Publication count over time

Table 3.1 Characteristics of included articles

Study Characteristic	Count	Sample	Count
Teaching Context		USA	29
Preservice	50	Turkey	14
Inservice	37	Australia	13
Pre & Inservice	5	Spain	11
		Canada	2
Schooling Level		Israel	2
Elementary/EC	37	Sweden	2
Middle School	5	Slovakia	2
High School	28	Argentina	1
Mixed Level	12	Bhutan	1
N/A	10	Brazil	1
		Chile	1
Study Design		Ghana	1
Qual (12 multi method)	47	Liberia	1
Quant	25	Netherlands	1
Mixed Method	4	Palestine	1
Qual + Quant	16	South Africa	1
		South Korea	1
		Southern Europe	1
		Sweden/Finland/Norway	1
		USA/South Korea	1
		Thailand	1
		Chile	1
		Indonesia	1
		Total	92

rightly question the kinds of interpretations one might make if Bhutanese ideas about emotions had been used. Recent research and theory of emotions that draws on constructionism and neuroscience suggests that the question is open about the extent to which it is relevant to interpret the emotions of one cultural group through the language labels, cultural assumptions, and meanings from another group (cf. Feldman-Barrett, 2017). This question applies not only across the Western/Eastern cultural groups, but also within Western countries. For example, the labelling of emotions in northern European countries differs somewhat from the labelling used in the United Kingdom and USA.

There is a slightly larger number of studies focusing on preservice teachers (50) rather than inservice teachers (37), and only 5 studies included both groups. Early childhood and elementary schooling are represented more strongly than other schooling levels, followed by high school, and then a mixture of levels. Although most studies clearly indicated the schooling level and focus on preservice teachers, inservice teachers, or both, we had to infer the schooling level in some cases due to a lack of specificity. A clear improvement for some authors would be to specify the study context more clearly to aid transferability of findings.

Qualitative study designs are more common than quantitative, and several qualitative studies adopted multiple methods. Only 4 studies adopted a mixed-methods design, and 5 studies included quantitative and qualitative data without using a mixed-methods research design.

Given the emphasis on elementary science teacher samples and qualitative designs, there is scope for future research on preservice and inservice teachers' emotions to adopt quantitative, mixed-method, or multi-method designs across all schooling levels. However, the overall number of studies (92) is minor compared to well-established research traditions in science education (e.g., conceptual change, socio-scientific issues), and the field would benefit from more research of any design, with any participant sample. Comparative studies across countries would offer benefit as only two included studies contained samples from different countries.

3.1 General Theoretical Trends Based on Discipline

As noted in Chap. 1, a diverse range of emotion theories from a range of disciplines informs science education research and education research more broadly. Table 3.2 presents an overview of the theoretical approaches used in the included studies that span the use of specific emotion theories and atheoretical studies (i.e., N/A in discipline and theory columns). As can be seen, psychological theories are dominant representing 37% of all studies, which is followed by atheoretical approaches with 32%. Of the psychological theories, general psychology, as opposed to more specific approaches (e.g., cognitive psychology, social constructionism) are most frequent. This is closely followed by psychological theories of self-efficacy, which include emotions as a key construct. It is not surprising to find that psychological perspectives or theories of emotion dominate the studies included. The rise of psychology in the nineteenth century, and its appropriation of emotions as a key topic of theorization and investigation during that time pervades popular and scholarly understandings of emotions.

In Table 3.2, Multi-disciplinary approaches refer to studies that integrated theories from different disciplinary origins, such as combining sociological and psychological theories. The category *Education* is used for studies that refer to literature on emotion from educational perspectives, but with no theoretical foundation in emotion theory, and for this reason, the additional label *atheoretical* is included. An illustrative example of such studies is those from the socio-scientific issues (SSI) tradition. In such studies, authors of included studies draw on the conceptualization of SSI from key SSI scholars. Authors also adopt ideas *about emotions* from key SSI scholars that are not grounded in emotion theories. In a similar way,

Table 3.2 Theoretical perspectives on emotion in science teacher education research

Discipline	Theory	Count	% Theory	Discipline % of total
Psychology	Psychology general	6	6	37
	Psychology/ Self-efficacy	5	5	
	Control Value Theory	4	4	
	Identity	3	3	
	Psychology/Education	2	2	
	Social constructionism	2	2	
	Affect	1	1	
	Appraisal theory/ Identity	1	1	
	Constructionist	1	1	
	Psychology/Discourse	1	1	
	Epistemic emotions/ affect	2	2	
	Psychology/Informal reasoning	1	1	
	Psychology/Linguistic	1	1	
	Psychology / Neuroscience	1	1	
	Psychology/ Behaviorist	1	1	
	Psychology/Biology	1	1	
	Emotional intelligence	1	1	
N/A	N/A	20	22	32
	Self-efficacy	4	4	
	Identity/ post-structuralism	2	2	
	Behaviorist/ Evolutionary	1	1	
	Communication	1	1	
	Discursive	1	1	
Sociology	Sociology	8	9	9
Multi-disciplinary	Sociology/ Psychology/Cognitive science	6	6	12
	Sociology/Psychology	4	4	
	Education/Cultural/ Feminist perspectives	1	1	
Education (atheoretical)	Educational perspectives	4	4	6

(continued)

Table 3.2 (continued)

Discipline	Theory	Count	% Theory	Discipline % of total
	Socio-scientific issues	2	2	
Philosophy/Ethics	Moral reasoning/Ethic of care	2	2	3
	Aesthetics	1	1	
Neuroscience	Neurobiology	1	1	1
Total		92	100	100

for studies classified under Educational (atheoretical), study authors draw on ideas of education scholars such as Noddings (1996) and Hargreaves (1998), who wrote about emotion in teaching, without reference to emotion theory or without proposing conceptualizations or theory of emotion.

With respect to the categories listed as N/A for discipline, this refers to atheoretical studies that mention emotions based on lay perspectives. Within the N/A group, there are also those studies that have no disciplinary basis associated with emotion, but they do adopt theories related to other constructs (e.g., self-efficacy, identity). In this sub-group of the N/A category, ideas about emotion used in the studies derive predominantly from the other constructs such as theories of self-efficacy and identity, but do not entail a specific theory of emotion.

After psychology, the most represented theoretical perspective on the emotions in preservice and inservice science teacher education comes from sociology. This has largely been through the application of interaction ritual theory and sociology of emotions in studies of science teacher emotion. This consistent body of work has spanned science students and science teachers' emotions with a focus on early career teachers. A broader range of sociological perspectives is available and could form the grounds for future research.

Despite psychology's dominant influence on studies of science teacher emotions, the breakdown of specific perspectives from the field offers scope for diversifying the focus even within this theoretical tradition. A strong tradition of teacher identity research is found in science education and education research more broadly. This can explain why this is the most frequently occurring topic among studies informed by psychological perspectives. Any one of the topics listed in Table 3.2 with a count of one could be expanded and provide necessary information that broadens our field's understanding of the emotions in science teaching. Epistemic emotions have enjoyed recent attention in the context of school children's science learning. It would be valuable to understand science teachers' epistemic emotions to complement the knowledge we have begun to develop about children and youth. Having offered a broad analysis of included articles, we now provide a detailed analysis by major topics of research.

As seen in Table 3.2, there is a range of theoretical options that remain under-researched that could form the foundation for future studies of preservice and inservice science teaching. One of the more intriguing offerings of late is the history of emotions field (cf. Dixon, 2012). Because of the lack of affiliation with any one disciplinary tradition (i.e., psychology, philosophy, or sociology), the history of emotions offers an expansive enterprise with revealing understandings of the evolution of thought about the emotions over time (see Dixon, 2012). An interesting under-representation among the disciplinary and theoretical perspectives is the philosophy of emotions (cf. Scarantino, 2016). There is a long and rich history of thought in philosophy forming a foundation of ideas underpinning psychology and sociology, as well as alternative perspectives, that can enrich future research on preservice and inservice science teachers' emotions. Given the prevalence of textual data in much of educational research, there is also more scope for adopting Kövecses (2003) emotion metaphor as a key construct. Michalinos Zembylas, in the works reviewed in thematic chapters, is one scholar who has addressed metaphor, but this represents a small proportion of all included research. Other obvious omissions are macrosocial emotion theories, which can deal with emotion at wider social scales than individual and small groups (e.g., Barbalet, 2001; Kemper, 2006; Turner 2007). Given the lasting impact of constructivism on education research, it is also surprising to see social psychology is largely absent (e.g., van Kleef, 2016).

3.2 Trends in Methods Used to Study Science Teacher Emotions

An analysis of methods and methodological approaches (see Table 2.2) reveals that qualitative research designs dominate the field, comprising 47 studies (51%) of the articles reviewed. Quantitative approaches account for 25 studies (27%), while mixed methods designs represent only 4 studies (4%). Additionally, 16 studies (17%) employed both quantitative and qualitative data collection without adopting a formal mixed-methods framework (labeled as Quant + Qual in Table 2.2). Among qualitative studies, interviews (I) emerge as the most prevalent data collection method, appearing in 31 studies, followed by observations (O) in 20 studies, and written artifacts (W) such as reflective journals or lesson plans in 15 studies. Open-ended surveys (So) and autoethnography (A) were employed in 11 studies each, whilst focus groups (FG) appeared in 7 studies. Within quantitative research designs, Likert-scale surveys (SL) predominate, utilized in 34 studies, and experimental designs (E) occurring in only 3 studies. This methodological landscape suggests that researchers investigating science teacher emotions favor qualitative contextualized approaches to capture emotional experiences, though the limited use of formal mixed-methods designs, as defined by Creswell (2021) indicates potential opportunities for more integrative research approaches that could strengthen this emerging field.

3.3 Topics of Investigation in Science Teacher Emotion Research

Through thematic analysis of topics forming the foci of empirical studies, we construct 6 themes to group included articles:

1. Classroom emotional climate ($n = 4$; 4.3%)
2. Science inquiry, laboratory, and practical work ($n = 17$; 18.5%)
3. Controversial issues ($n = 15$; 16.3%)
4. Technology ($n = 8$; 8.7%)
5. Science teachers' self-efficacy and identity ($n = 19$; 20.7%)
6. Science teachers' work ($n = 29$; 31.5%).

The number and proportion of included studies associated with the themes, as shown in parentheses, indicates areas of research concentration with studies of emotion relating to science teachers' work being the most prevalent. Such focus coincides with early interest in educational research on the emotions associated with teachers' work conditions (cf. Hargreaves, 1998; Noddings, 1996). Science inquiry, laboratory work, and practical work, controversial issues, and teacher self-efficacy and identity were almost equal in being the second most prevalent theme, whereas classroom emotional climate and technology were almost equal as least represented themes. We anticipate the rise of general artificial intelligence in educational contexts may turn more attention to the technology theme in the future. It was often the case that a focus on emotions arose in studies about self-efficacy and identity due to the presence of the construct in theories of self-efficacy and identity, rather than serving as the central focus of that theme. This pattern was also observed within most of the themes, and in some cases, it was not associated with any theory, but seemed to represent everyday ideas about emotions, which were incidental finds in studies focused on the thematic topic.

The list of themes, whilst inclusive of several focal areas for science education research, holds obvious omissions (cf. Lin et al., 2025; Odden, et al., 2021; Tosun, 2024). For example, key topics of science education research including conceptual change, analogy, teaching metaphors, metacognition, representations, and activism are not represented in the included studies. Research on emotions associated with this range of very established and more recent fields of inquiry offers an immediate angle for future research. Overall, due to the relatively small number of studies across the board, each topic represented in our theme names would benefit from more research.

Within most themes, we found limited connectivity between studies, leading to our discussion of key aspects of these works independently in Chaps. 4–9. This fact directs attention to future research which focuses on building broader and deeper knowledge connected to existing studies. Such an approach would offer a stronger evidence base for informing further research, practice, and policy, than what is currently available. In addition, better conceptual connectivity between studies would support advancement of knowledge through cohesion of prior and new work. Diversity in theory and method must be maintained, but greater synergy seems to be the most pressing need.

In Chaps. 4–9, we present our review of included studies beginning with theoretical trends and followed by empirical findings associated with each theme. We offer an overview of the key study outcomes; synthesize findings; when possible, offer critique; and suggest future directions and implications for each theme. Our focus is exclusively on reporting the emotional outcomes presented in the included studies and not on other outcomes presented by the authors.

References

NB Articles included in the systematic review are indicated using an asterisk (*)

Barbalet, J. (2001). *Emotion, social theory, and social structure: A macrosociological approach.* Cambridge University Press.

Creswell, J. W. (2021). *A concise introduction to mixed methods research.* Sage Publications.

Dixon, T. (2012). *From passions to emotions: The creation of a secular psychological category.* Cambridge University Press.

Feldman-Barrett, L. (2017). *How emotions are made: The secret life of the brain.* Macmillan.

Hargreaves, A. (1998). The emotional politics of teaching and teacher development: With implications for educational leadership. *International Journal of Leadership in Education, 1*(4), 315–336.

Kemper, T. D. (2006). Power and status and the power-status theory of emotions. In J..E. Stets & J. H. Turner (Eds.), *Handbook of the sociology of emotions.* Springer. https://doi.org/10.1007/978-0-387-30715-2_5

Kövecses, Z. (2003). *Metaphor and emotion: Language, culture, and body in human feeling.* Cambridge University Press.

Lin, T.-J., Lin, T.-C., Potvin, P., & Tsai, C.-C. (2025). Research trends in science education from 2018 to 2022: A systematic content analysis of publications in selected journals. *International Journal of Science Education, 47*(4), 510–538. https://doi.org/10.1080/09500693.2024.2340811

Noddings, N. (1996). Stories and affect in teacher education. *Cambridge Journal of Education, 26*(3), 435–447.

Odden, T. O., B., Marin. A., & Rudolph, J. L. (2021). How has *Science Education* changed over the last 100 years? An analysis using natural language processing. *Science Education, 105,* 653–680. https://doi.org/10.1002/sce.21623

*Rinchen, S., Ritchie, S. M., & Bellocchi, A. (2016). Emotional climate of a pre-service science teacher education class in Bhutan. *Cultural Studies of Science Education, 11,* 1–26. https://doi.org/10.1007/s11422-014-9658-0

Scarantino, A. (2016). The philosophy of emotions and is impact on affective science. In L. F., Barrett, M., Lewis, & J. M. Haviland-Jones, (Eds.), *Handbook of emotions* (4th ed., pp. 3–48). The Guilford Press.

Tosun, C. (2024). Analysis of the last 40 years of science education research via bibliometric methods. *Science & Education, 33,* 451–480. https://doi.org/10.1007/s11191-022-00400-9

Turner, J. H. (2007). *Human emotions: A sociological theory.* Routledge.

van Kleef, G. A. (2016). *The interpersonal dynamics of emotion: Toward an integrative theory of emotions as social information.* Cambridge University Press.

Chapter 4
Emotional Climate of Science Education Classes

A body of research on emotions in science teacher education focuses on collective aspects of emotion called *emotional climate* (Bellocchi et al., 2013; Bellocchi et al., 2014; Rinchen, et al., 2016; Tobin et al., 2013). These studies address international contexts including secondary science teacher education in Australia (Bellocchi et al., 2013, 2014), elementary science education in Australia (Tobin et al., 2013), and Bhutanese secondary physics teacher education (Rinchen et al., 2016). In addition to the broader focus on collective emotional climates (EC) these studies share theoretical foundations in Randall Collins' (2004) interaction ritual theory and Turner's (2007) sociological theory of emotions.

Common to all studies is the focus on in-the-moment social interactions and emotions, and their mediation of EC. Tobin et al. (2013) explored the social interactions that gave rise to positive and negative EC in relation to teaching and learning in a Year 7 science class, by analyzing the interactions of a teacher and her students. Bellocchi et al. (2013) investigated how salient social interactions are related to EC in preservice science teacher education classes that featured debates about the SSIs. In a later study, Bellocchi et al. (2014) investigated the EC of a preservice secondary science education class to reveal practices that produce and reproduce high quality learning experiences. Exploring a distinct cultural context for EC research, Rinchen et al. (2016) investigate EC in Bhutan. Here, the nation of Bhutan and its unique macro-social policy of *Gross National Happiness*, which seeks to enhance the nation's emotional wellbeing and prosperity, served as the context for examining preservice science teachers' perceptions with a view to enhance their professional development.

A. Bellocchi et al., *Emotion Research in Science Teacher Education*, Sociocultural Explorations of Science Education 32, https://doi.org/10.1007/978-3-032-07711-0_4

4.1 Theories of Emotion in Emotional Climate Research

Sociological theories, including Turner's (2007) theory of emotions, have been adopted in all studies of EC in science education. Bellocchi et al. (2013), Rinchen et al. (2016), and Tobin et al. (2013) also draw on interaction ritual theory (Collins, 2004), as they focus on the social interactions between preservice teachers or between students and their teachers that mediate EC in particular settings. Tobin et al. (2013) describe their study as multi-theoretic, drawing on a range of sociocultural frameworks to understand the EC of a Year 7 science. More specifically, they examine how social artifacts, including the practices of students and teachers, are inscribed with emotional energy (Durkheim, 1915/1965), which in turn reflects the EC of classroom events as the science curriculum is enacted. Emotional energy is a constant steady flux of emotion which differs from discrete emotions (e.g., anger, fear, joy etc.).

4.2 Empirical Outcomes of Emotional Climate Research

The study of science classroom EC was initiated by Tobin et al. (2013), with the experiences of an elementary inservice science teacher. Employing multiple methods in their event- orientated social analysis, the researchers observed 11 lessons over a 5-week period and video-recorded seven of them. The researchers rated the classroom climate at 3-min intervals on a five-point scale (1-very negative, 2-negative, 3-neutral, 4-positive, 5-very positive) using audience response technology. The results showed that EC varied continuously throughout a lesson. Interactions between the teacher and students that gave rise to positive EC and enabled learning involved humor, fluent dialogic exchanges, and *collective effervescence* (characterized by, for example, laughter and hand clapping). In contrast, high levels of frustration and negative EC arose when the teacher and/or students sought to establish and maintain control over the class by restricting the participation of others to spectator roles, which was detrimental to teaching and learning.

One of the key contributions this study made to science education research was to establish methods for EC measurement. The researchers used audience response technology with interval-based recording of EC based on a five-point scale in combination with co-generative dialogues. In cogenerative dialogues, the teacher and small student groups reflected on short episodes of classroom interaction to identify ways of improving negative experiences in future lessons. The study also included video data, which enabled conversation analysis, prosody analysis (i.e., the rhythm and intonation of speech), and analysis of nonverbal action (i.e., facial expressions, body movements, and gestures). A limitation of Tobin et al.'s study was that the EC ratings were based on the researchers' perspective, leaving student perceptions of EC unaddressed, providing a foundation for subsequent research using similar methods in preservice teacher education in Australia and Bhutan.

Bellocchi et al. (2013) explored EC in the context of a course focused on teaching Australian preservice science teachers to use debates about SSIs (nuclear energy alternatives, climate change, and genetically modified foods) to engage secondary students with science. More specifically, they were interested in understanding what micro-processes of interactions were associated with positively and negatively valenced EC, and the extent to which formal and informal class rituals related to the valence and intensity of EC. In this multi-method, event-orientated study, 14 preservice science teacher participants used audience response technology to rate their perceptions of EC at 3-min intervals on a five-point scale (cf. Tobin et al., 2013). Using averages of the participants' EC ratings over time, graphical representations of the EC variations were used to identify key classroom events, in video data, associated with shifts from positive/very positive EC (i.e., *peaks*) to negative/very negative EC (i.e., *troughs*). EC data were supplemented by analysis of field notes recorded during and after the debate lesson, conversations recorded in salient video clips, and non-verbal behaviors (i.e., prosody, body movements and facial expressions).

Bellocchi et al. (2013) found that engaging debate presentations and natural classroom interaction rituals were associated with positive EC, whereas formal classroom rituals were associated with decreases in EC, or troughs in the EC graph. There were no appreciable differences in classroom EC associated with the different debate topics, suggesting that preservice teachers' perceptions of EC were directed at other aspects of the interactions, which included the extent to which the interactions invited their active participation in class activity. For example, when teams presented their debate arguments, the audience was positioned to listen to the argument with no scope for interactive participation. In this case, the EC ratings corresponded with the level of excitement and engagement that the debaters fostered during their presentations.

A key contribution of the study by Bellocchi et al. (2013) to preservice science teacher education is its identification of the micro-processes that underpin successful and unsuccessful class interactions associated with positive and neutral EC. The implication of this finding for practice is significant because it can support educators to improve EC through appropriate changes to prosody, gestures, and classroom discourse. The study also refined the EC construct by proposing the notion of *intensity of EC*, noting that valance (positive/negative) alone was not sufficient to explain all the researcher's empirical observations when class EC was positive, most of the time.

In a second Australian study of EC in preservice science teacher education, Bellocchi et al. (2014) investigated associations between individual emotional expressions with classroom EC and perceptions of high-quality science teacher education. Video recordings of classroom interactions were analyzed as were recordings of co-generative dialogues, or cogens. Cogens, which involved group reflections between researcher, teacher educator, and 4–5 preservice teachers, were adopted to reveal preservice teachers' perceptions of high-quality classroom experiences. Twenty-eight preservice teachers participated in the study by rating EC as described above for Bellocchi et al.'s (2013) study.

This study focused on the emergence of two kinds of science demonstrations that were identified as high-quality learning experiences by participating students: predict-observe-explain (POE) demonstrations of discrepant events, and role-plays.

A general pattern of EC ratings showed that during lectures and other forms of formal discussion about science concepts, the EC ratings were in the low negative to slightly positive range. Higher positive ratings were consistently observed with science demonstrations. During and following the demonstration, the teacher educator typically shared her reflections on the use of such approaches for instruction and described some practical situations she had encountered.

Preservice teacher EC ratings were highly positive during the demonstrations, while lower positive ratings were evident during the teacher educator's reflections about using demonstrations to teach science. Despite these lower ratings, the preservice teachers explained during co-generative dialogues that they considered the teacher educator's in-the-moment reflections as high-quality learning experiences because these reflections externalized aspects of their practice, which supported them to reflect on the considerations they would need to make when implementing demonstrations in their future science classes.

Another contribution the study made, extending Bellocchi et al.'s (2013) work, is to connect individual preservice teacher emotions with EC ratings. In 2013, Bellocchi et al. did not analyze video data extensively to establish connections between individual emotions and EC. In the 2014 study, this link was made through the analysis of facial expression. This methodological contribution revealed that high positive EC was associated with expressions consistent with experiences of individual positive (i.e., pleasant) emotions. Another association made in that study, as with Bellocchi et al.'s (2013) work, was that higher positive EC ratings occurred when classroom interactions followed dialogic patterns (i.e., where different speakers contribute to the construction of classroom topics), and lower EC ratings corresponded with univocal interactions such as the teacher educator's lectures or long periods of talk. Although univocal interactions received lower climate ratings, preservice science teachers considered univocal talk valuable if it provided them with practical ideas for classroom practices.

Collectively, the two EC studies by Bellocchi et al. (2013, 2014) illustrate that although EC can slump to low positive values, preservice teachers may still value activities associated with less positive EC. What is important, as both studies note, is that although low EC lesson segments may be less emotionally motivating, they exist in a context where more motivating experiences that are rated with high positive EC are also evident. In both studies, the authors caution that to assume that low positive EC ratings could be maintained without the incidence of higher positive EC segments would be misguided.

Although both studies by Bellocchi et al. (2013, 2014) focus on the role EC play in preservice science teacher education, neither makes connections between the classroom and broader social structures, or macro-social phenomena (e.g., policy). This limitation was addressed in a study of Bhutanese preservice physics teachers by Rinchen et al. (2016), who explored connections between Bhutan's *Gross National Happiness* policy and classroom EC and emotions. The study considered the ways in which emotions can operate across macro- (e.g., educational system; policy), meso- (e.g., university, schools, or classroom cultures), and micro-social levels (i.e., face-to-face interaction).

The aim of this ethnographic case study conducted over a 5-month period was to investigate how preservice science teachers' professional development can be enhanced, such that they support the realization of the *Gross National Happiness* policy by effectively producing positive ECs in school classrooms. Multi-methods including video and audio recordings of teaching episodes, preservice teacher EC ratings at 3-min intervals (cf. Bellocchi et al., 2013), teacher and student interviews, and researcher field notes were adopted.

Noting the common educational experience whereby Bhutanese students are submissive to their teachers and often deliver choral answers to teacher questions, Rinchen et al. (2016) reported that the use of student-led presentations using video clips and models, co- teaching, and interactive whole-class discussions were associated with positive EC and a classroom structure that went against the dominant submissive model of student–teacher interaction. Individual engagement, frequent interactions, informality of interaction, and the use of learning resources were identified as enabling conditions for sustaining positive EC. Conversely, negative, or low positive EC resulted when student presenters were unprepared and delivered lower quality classroom presentations, which gave rise to emotions such as frustration and disappointment. The study also confirmed the principles of emotional arousal as reported by Bellocchi et al. (2013, 2014), which were conducted in Western contexts (e.g., lower positive EC or negative EC were associated with lectures).

An important contribution made by Rinchen et al. (2016) to science education research was to identify the need for individuals to access relevant learning resources to transform classroom practices. The presence of appropriate resources produced and sustained mutually focused interactions that engaged preservice teachers and fostered positive EC. This finding extends the enabling conditions for positive EC originally proposed by Metiu and Rothbard (2012). The authors assert that the reproduction of the structures associated with positive EC in their study was important for supporting realization of the *Gross National Happiness* policy, particularly given the importance of quality science education to building human capital in a developing country like Bhutan.

Use of related or the same theories of emotion across these studies presents a benefit in the conceptual alignment offered, which allows alignment of findings across studies. However, even when studies adopt emotion theories, it is important to note that theory narrows insights to the specific emotion concepts and constructs represented within the theory. A clear omission from a theoretical stance across these studies is sensitivity to the impact of socio-cultural factors (e.g., class, race, gender identity) on observed emotions and their interplay with science teacher education. Critical theories of emotion would bring such sensibility to research in this, and numerous other, themes we will encounter in the book.

4.3 Conclusion

Research into EC in science teacher education has advanced our understanding of the collective and dynamic nature of emotional experiences within science teacher education. Grounded in sociological theories of emotion (Collins, 2004; Durkheim, 1915/1965; Turner, 2007), these studies explore how in-the-moment social interactions mediate collective emotional experiences in preservice science teacher education contexts (Bellocchi et al., 2013, 2014; Rinchen, et al., 2016), or in classrooms as the science curriculum is enacted (Tobin et al., 2013).

Methodologically, the reviewed studies adopt event-oriented, multi-method approaches that integrate quantitative measures (e.g., interval-based EC ratings using audience response technology) with qualitative data (e.g., co-generative dialogues, video analysis, field notes). Conversation analysis, prosody analysis and analysis of nonverbal behavior has also been enabled through video recordings of classroom interactions. These methods support in-depth, fine-grained analysis of the emotional ebb and flow of classroom interactions by identifying salient events that give rise to EC. Although Tobin et al.'s (2013) study made a key contribution to science education research on EC by establishing methods for its measurement, EC ratings were based on the researchers' perspective; a limitation that was addressed by the subsequent studies conducted in preservice teacher education contexts that captured student perceptions of EC.

Empirically, the studies demonstrate that positive EC is linked with interactive, participatory, and dialogic teaching practices—including high-quality learning experiences such as demonstrations of discrepant events, and role-plays—while univocal instruction and formal rituals correlate with neutral or negative EC. Importantly, these findings show that preservice teachers still value learning experiences associated with lower EC when they are encountered alongside more motivating experiences that are rated with high positive EC.

Key theoretical advancements offered by two of the reviewed studies are the distinction between valence and intensity in understanding EC (Bellocchi et al., 2013), as well as the linkage between individual emotional expressions and collective EC ratings (Bellocchi et al., 2014).

Rinchen et al.'s (2016) study further expands the field by examining EC through a macro–meso–micro lens, demonstrating how Bhutan's *Gross National Happiness policy* shapes classroom emotional dynamics by making connections to broader social structures and macro-social phenomena (in this case, policy). This ethnographic work emphasizes the importance of learning resources to transform classroom practices and foster positive EC, while the reproduction of the structures associated with positive EC was important to supporting the realization of Bhutan's *Gross National Happiness* policy.

Following Rinchen et al.'s (2016) study, there has been a paucity of research that has explored the EC in the context of preservice or inservice science teacher education. Noting that Tobin et al.'s (2013) study was conducted more than a decade ago, further inquiry into the ECs of science classrooms in warranted, particularly in the context of current complexities associated with, for example, out of field teachers in science (cf. Campbell et al., 2024; Hobbs et al., 2022; Luft et al., 2020), high-stakes assessment in senior secondary school science (e.g., Cumming, 2020), the expansion of online learning, the continued concerns around students' waning engagement with school science, and links to teacher preparation and teaching quality (e.g., Steidtmann et al., 2023). A general limitation of studies on emotional climate and science teacher education is that socio-cultural factors that might be important for explaining emotional responses to classroom experiences are not a central feature of many of the studies. Future research should consider how science teachers' background may shape their experiences of classroom climates and how they may shape the climates they produce in their classes.

References

NB Articles included in the systematic review are indicated using an asterisk (*)

*Bellocchi, A., Ritchie, S. M., Tobin, K., Sandhu, M., & Sandhu, S. (2013). Exploring emotional climate in pre-service science teacher education. *Cultural Studies in Science Education, 8*, 529–552. https://doi.org/10.1007/s11422-013-9526-3

*Bellocchi, A., Ritchie, S. M., Tobin, K., King, D., Sandhu, M., & Henderson, S. (2014). Emotional climate and high quality learning experiences in science teacher education. *Journal of Research in Science Teaching, 51*, 1301–1325. https://doi.org/10.1002/tea.21170

Campbell, C., Vale, C., & Speldewinde, C. (2024). Teaching science out-of-field: Beliefs and practices. *European Journal of Mathematics and Science Education, 4*(2), 133–148.

Collins, R. (2004). *Interaction ritual chains*. Princeton University Press.

Cumming, J. J. (2020). Senior secondary school assessment and standard-setting in Queensland, Australia: Social context and paradigmatic change. *Assessment in Education: Principles, Policy & Practice, 27*(2), 160–177.

Durkheim, E. (1915/1965). *The elementary forms of the religious life* (Trans. J. W. Swain). Allen & Unwin.

Hobbs, L., Du Plessis, A.E., Oates, G., Caldis, S., McKnight, L., Vale, C., O'Connor, M., Rochette, E., Watt, H., Weldon, R., Richardson, P., & Bateup, C. (2022). National summit on teaching out-of-field: Synthesis and recommendations for policy, practice and research. https://ooftas-collective.squarespace.com/s/TOOF-National-Summit-Report.doc

Luft, J. A., Hanuscin, D., Hobbs, L., & Törner, G. (2020). Out-of-field teaching in science: An overlooked problem. *Journal of Science Teacher Education, 31*(7), 719–724.

Metiu, A., & Rothbard, N. P. (2012). Task bubbles, artefacts, shared emotion, and mutual focus of attention: A comparative study of the micro processes of group engagement. *Organization Science, 24*, 455–475. https://doi.org/10.1287/orsc.1120.0738

*Rinchen, S., Ritchie, S. M., & Bellocchi, A. (2016). Emotional climate of a pre-service science teacher education class in Bhutan. *Cultural Studies of Science Education, 11*, 1–26. https://doi.org/10.1007/s11422-014-9658-0

Steidtmann, L., Kleickmann, T., & Steffensky, M. (2023). Declining interest in science in lower secondary school classes: Quasi-experimental and longitudinal evidence on the role of teaching and teaching quality. *Journal of Research in Science Teaching, 60*(1), 164–195.

*Tobin, K., Ritchie, S. M., Oakley, J., Mergard, V., & Hudson, P. B. (2013). Relationship between emotional climate and the fluency of classroom interactions. *Learning Environments Research, 16*, 71–89. https://doi.org/10.1007/s10984-013-9125-y

Turner, J. H. (2007). *Human emotions: A sociological theory*. Routledge.

Chapter 5
Emotion and Science Inquiry, Laboratory and Practical Work

Following the emphasis on experiential learning deriving from Dewey (1910), science laboratory, practical work, and scientific inquiry (or student investigations) have captured the attention of curriculum reform, teaching, and research for considerable time. The range of studies of inservice and preservice teachers' emotions relating to science inquiry chiefly concern their emotional responses (positive and negative) to participating in science inquiry (Adler et al., 2019; Alarcón-Orozco, et al., 2025; Davidson et al., 2020; Dreon & McDonald, 2012; García-Ruiz, et al., 2022; Jaber et al., 2021; López-Banet, et al., 2021; Melville & Bartley, 2013; Ritchie, et al., 2013; Ritchie & Newlands, 2017). Several studies focus on participants' emotional responses to specific inquiry approaches, such as demonstrations (Bellocchi et al., 2018a), place-based approaches (Cook & Buck, 2017), dissections (Fančovičová, et al., 2013), and the use of terrestrial isopods and spiders (Wagler & Wagler, 2017, 2022). Teacher identity is the focus of Meville and Bartley's study (2013), whereas Gilbert and Byers (2017) examine wonder as a tool for engaging elementary preservice teachers in learning about science inquiry, and Bahcivan (2019) examines the relationships between preservice teachers' epistemic beliefs, epistemic emotions, and argumentation. Although this study does not strictly concern science inquiry, we align it with this theme, given the importance of teachers' epistemic beliefs (i.e., ideas about knowledge and knowing) to their engagement with and enactment of science inquiry; an epistemic activity. Overall, 17 studies are included in this theme, representing 18.3% of included studies.

A. Bellocchi et al., *Emotion Research in Science Teacher Education*, Sociocultural Explorations of Science Education 32, https://doi.org/10.1007/978-3-032-07711-0_5

5.1 Theories of Emotion in Science Inquiry and Laboratory and Practical Work Research

Studies addressing emotion and inquiry and laboratory or practical work range from the atheoretical (Adler, et al., 2019; Alarcón-Orozco, et al., 2025; Gilbert & Byers, 2017), to those adopting psychological or biological perspectives (Bahcivan, 2019; Fančovičová, et al., 2013; López-Banet et al., 2021; Wagler & Wagler, 2017, 2022), sociological perspectives (Bellocchi et al., 2018a; Cook & Buck, 2017; Ritchie et al., 2013; Ritchie & Newlands, 2017), and post-structuralist perspectives (Melville & Bartley, 2013). Although Davidson et al. (2020) acknowledge three key conceptualizations of emotions in the literature (i.e., individualistic, sociocultural, and postmodernist), they privilege a postmodernist view, acknowledging that although affect can be shaped by individualized experiences and sociocultural contexts, it can be expressed outwardly and experienced inwardly, and recognized both in-the-moment and upon reflection. Jaber et al. (2021) refer to *epistemic empathy*; that is, when teachers draw upon their own experience of engaging in inquiry to connect to students' epistemic experiences. Dreon and McDonald (2012) draw on Pintrich et al.'s (1993) work on affect and its relation to conceptual change, once again without reference to primary emotions literature. The study's phenomenological approach supports the atheoretical stance taken towards emotion. In contrast, García-Ruíz et al. (2022), who also take an atheoretical approach to investigate emotions, do not present a study design suited to this approach. The study adopted common emotion labels, possibly from everyday language, into a questionnaire. Alarcón-Orozco, et al. (2025), in contrast, discuss a range of emotion theories ranging from disciplinary origins such as psychology, educational psychology, and neuroscience. Emotion labels and constructs aligned with control-value theory then inform the data collection methods and analyses (cf. Pekrun, 2006).

5.2 Empirical Outcomes on Emotion in Science Inquiry, and Laboratory and Practical Work Research

Empirical outcomes are grouped and presented around two main ideas: (1) aversion emotions in biology practical work ($n = 3$); and (2) emotions when engaging in or supporting science inquiry ($n = 14$).

5.2.1 Aversion Emotions in Biology Practical Work

The three studies investigating preservice science teachers' aversion emotions–disgust and fear–focus on practical aspects of employing dissections (Fančovičová et al., 2013) and terrestrial isopods and spiders (Wagler & Wagler, 2017, 2022) for science instruction. Overall, the studies report that prior experience with practical work of this nature tends to increase preservice teachers' desire to engage with dissection or animal species that that they would otherwise avoid.

Fančovičová et al. (2013) examined nearly 400 Slovakian preservice biology teachers perceived disgust and attitudes towards dissections. The authors found that participants who had prior experiences of dissections (live or virtual) reported lower levels of disgust and attributed greater importance to dissections in the science curriculum than those without any prior experiences. Female participants and those who reported owning animals at home demonstrated higher levels of disgust towards dissections.

Fančovičová et al. (2013) draw upon Pekrun et al.'s (2002) psychological approach to academic emotions, which connects emotions and academic outcomes, and Randler et al.'s (2005) environmental education study of student's disgust, which connects learner emotions to outcomes. Apart from their reference to academic emotions, theory is not used to make sense of observational data. Rather, the outcomes of other studies in education (Pekrun et al., 2002; Randler et al., 2005) are used to compare and justify the focus on learner's disgust during biology dissections. Lack of theorization of disgust, and emotions more generally, leads to conflation between attitudes and emotions. Although interesting outcomes are reported about the correlations between learner disgust and different approaches to dissections, including real and virtual dissections, learner disgust is inferred from participant responses to a questionnaire. Given that participants were university students, and the questionnaire included questions about past experiences with dissections in elementary school, the validity of reports after prolonged periods of time is questionable. Disgust is a well-researched emotion, particularly in relation to facial and gestural actions, further work could theorize disgust, make observations of students working with different kinds of dissections, and interpret what function disgust may have during dissections rather than presuming a priori that disgust experienced during dissections is undesirable.

Middle school preservice science teachers' fear and disgust reactions to spiders is the focus of Wagler and Wagler's (2017) study, which investigates how these emotions impact their instructional choices. The preservice teachers in the study are all non-science majors studying a science methods course in a Southwestern U.S. university. The study is informed by biological and evolutionary perspectives of emotion that define fear and disgust as protective emotions that benefit survival.

A pre/post randomized design with a control group was adopted, wherein preservice teachers viewed a live spider in a clear container and reported their levels of fear and disgust on a 5-point Likert scale. They also reported the likelihood of incorporating information about spiders, and the likelihood of incorporating information about the ecological role of spiders, into their future instruction, on a 5-point Likert scale. Throughout semester, the test group participated in hands-on activities about spiders, whereas the control group did not. At semester's end, a post-test was administered where their fear/disgust and likelihood of incorporating spiders in future instruction was reassessed.

Using pre-tests, Wagler and Wagler found that preservice teachers felt moderate levels of fear and disgust towards the spider, and that those who participated in hands-on activities reported lower levels of these emotions at post-test. The control group participants reported no change in their fear and disgust. A similar pattern followed with preservice teachers' intentions to incorporate information about spiders and information about their ecological roles with test-group participants reporting increased intentions to incorporate instruction and control-group participants reporting no change. A limitation of the analysis was that correlations between reports of fear and disgust and instructional intentions could not be made.

Subsequent work by the same authors examined the extent to which fear and disgust influenced preservice middle school science teachers' intentions to incorporate information about terrestrial isopods and their ecological services in their science classrooms (Wagler & Wagler, 2022). The authors employed the same research design and similar participants as the 2017 study. All participants viewed a live isopod in a clear container and reported moderate levels of fear and disgust at pre-test on a 5-point Likert scale. The test-group then engaged in a range of hands-on science inquiry activities about terrestrial isopods over the course of the semester, while the treatment group did not. Data analysis revealed a decrease in fear and disgust ratings for the test group, while no change was observed for the control-group. The test-group also expressed a strong desire to teach children about terrestrial isopods, whereas the control-group reported no interest in doing so.

Wagler and Wagler (2022) note that the educational intervention in their latter study was more effective than it was in the 2017 spider study, as the research participants reported lower levels of fear and disgust to begin with. They also demonstrated higher intentions to teach students about terrestrial isopods. The authors conclude that different species can elicit different attitudes and emotions, and that, based on the study's findings, terrestrial isopods should be the first living arthropod used in science teacher preparation programs with a view to enhance the likelihood that they will incorporate them into their classroom practice, particularly given that spiders elicit higher levels of fear. Like their 2018 study, correlations between fear, disgust, and instructional intentions were not made. The forced choice between two emotion labels presents a limitation as other emotions that were experienced by the participants remain unknown.

5.2.2 Emotions When Participating in or Supporting Science Inquiry

The role of preservice and inservice teachers' emotions as they engage in or support science inquiry instruction is the focus of 10 studies (i.e., Adler et al., 2019; Alarcón-Orozco, et al., 2025; Bellocchi et al., 2018a; Davidson et al., 2020; Dreon & McDonald, 2012; García-Ruiz, et al., 2022; Jaber et al., 2021; López-Banet et al., 2021; Ritchie et al., 2013; Ritchie & Newlands, 2017). Bellocchi et al. (2018a) examine how emotional energy arising from science demonstrations conducted with secondary preservice teachers as successful interactions can empower them to recreate the pleasant emotional energy that they experienced by re-enacting demonstrations in their own school classroom practice. Cook and Buck (2017) examine the affordances of place-based science inquiry for fostering emotional energy and solidarity. A key idea that underpins the central findings of these studies is that emotional trajectories across pleasant and unpleasant emotions shape preservice and inservice teachers' views of and/or teaching practices associated with practical activities and science inquiry.

Seven of the reviewed studies focus on the emotional experiences of preservice science teachers engaging in science inquiry during their initial teacher education programs (Alarcón-Orozco, et al., 2025; Bahcivan, 2019; Bellocchi et al., 2018a; García-Ruiz, et al., 2022; Gilbert & Byers, 2017; López-Banet et al., 2021; Ritchie & Newlands, 2017). When preservice teachers initially engage in science inquiry, a range of unpleasant emotions such as anxiety, fear, concern, worry, frustration, annoyance, and discomfort are reported. For example, a study by Ritchie and Newlands' (2017) examined the emotional arousal of two secondary preservice teachers engaged in the planning and enactment of an extended chemistry experimental investigation in their science methods course at an Australian university. In the initial investigation stages, participants reported frustration and annoyance with challenges in setting up apparatus or when the experiment went wrong; however, achieving success with the investigation produced intense happiness and resolved the unpleasant emotions. In this way, Ritchie and Newlands describe the inquiry as a 'transformative activity' that impacted on preservice teachers' subsequent science practices.

Adopting a psychological view of emotion that includes cognitive and behavioral dimensions, López-Banet et al. (2021) implemented and evaluated an instructional sequence about acids and bases, using a model-based inquiry (MBI) approach–wherein students generate, test, and devise their own scientific model–on Spanish secondary preservice teachers' emotional and cognitive engagement. Here, the authors sought to examine the relationship between participants perceived knowledge after engaging in the MBI sequence and their emotions. Two groups of postgraduate preservice teachers from diverse disciplinary backgrounds ($n = 27$; $n = 18$) participated in the study. None of the participants had any previous experience with MBI sequences. The 2-hour instructional sequence investigated the effects of chewing gum on the mouth's pH levels and was taught by two of the study's authors. Preservice teachers were required to respond individually to 10 tasks before discussing them with the whole class.

At the end of the sequence, a questionnaire, the *Knowledge and Prior Study Inventory*, was administered to gauge participants' self-regulation of learning and their emotions. The questionnaire asked participants to associate provided emotion labels with all key moments in the instructional sequence. Ten emotion labels were included: rejection, concentration, insecurity, interest, boredom, confidence, satisfaction, dissatisfaction, bashfulness, and surprise. Participants were also asked to rate how well they understood the science content before and after the sequence on a scale of 1–5 (1: I know nothing, to 5: I can explain it to a friend), for each task, in chronological order.

Quantitative data analysis revealed that the most salient emotions reported by the preservice teachers were interest, concentration, and surprise. Significant correlations were observed between knowledge and the emotions identified by the participants in all key activities. For some tasks, higher levels of confidence or satisfaction related positively to greater perceptions of learning, and preservice teachers who felt that they learned more expressed feels of security in relation to their science inquiry skills, such as developing an experimental design to solve a given problem. Higher levels of satisfaction and interest were also linked to participants' knowledge of model-based inquiry. Conversely, preservice teachers who felt that they did not learn as much expressed more unpleasant emotions in tasks that involved science inquiry skills.

While the authors assert that the study's findings indicate significant relationships between knowledge and emotions, it is important to highlight that data reflects participants' *perceived* knowledge. Additionally, a lack of detail regarding the MBI sequence makes it difficult to draw inferences about this inquiry approach and the reported findings. It is unclear what evidence-based model or explanation the preservice teachers sought to develop their participation in the MBI sequence; insufficient detail is provided about the nature of the 10 key tasks.

The focus of García-Ruiz, et al.'s (2022) empirical work was to identify change in Pre-service Science Teachers (PSSTs) inquiry-based instruction using mixed-methods design within a case study of five PSSTs enrolled in a Master of Education course at the University of Málaga, Spain. Two questionnaires with Likert items were used to explore relationships between perceptions of instruction, emotional and cognitive profiles, emotions, and inquiry design performance. Questionnaires were administered before and after tailored inquiry-based instruction. Measures were used to compare PSSTs' perceptions about relevant variables in their roles as University students and as (imagined) future classroom teachers. Our focus is on reporting findings relevant to emotions. When adopting a PSST student role in response to the provided instruction, PSSTs reported pleasant emotions. However, when the learning task positioned PSSTs as teachers, there was not directional relationship between experience of the task and emotional valence and there was no discernible relationship to attained levels of understanding about inquiry-based instruction. The authors found PSSTs' held insecurities when considering complex design steps of inquiry-based instruction such as curriculum application, management and assessment. The ad-hoc use of an emotion questionnaire poses limitations for this study as the phenomena under investigation, emotions, cognition, inquiry design performance are complex. Emotion theory that could address connections to cognition or learning would have supported interpretation of study findings.

Gilbert and Byers (2017) explore the affordances of wonder as a pedagogical tool in elementary preservice science teacher education in Australia. They argue that while the link between wonder and science is *writ large* throughout history, wonder is rarely utilized as an explicit tool in initial teacher education. The authors draw upon the notion of epistemic affect to postulate whether purposefully supporting preservice teachers to connect wonder, emotion, aesthetics, and science learning would shape their perceptions of science and their future practice.

Utilizing an instrumental qualitative case study design, the experiences of 23 female preservice teachers enrolled in a primary education program at a metropolitan university were examined during their final 18 months of study. In the first semester, wonder was used to connect the participants with science content, while the second semester was focused on their development of inquiry-based units for enactment during their 3-week professional experience placement. Science experiences usually began with a discrepant event, or some other way of invoking wonder about common phenomena, with a view to fostering an emotive connection with science. The preservice teachers kept journals to record their wonders, questions, and vexations, and engaged in group discussion before exploring key science content. Other pedagogical approaches were employed to encourage preservice teachers to wonder about science phenomena related to the primary science curriculum, including wonder lists, wonder walls, and wonder projects based on their individual interests. Artefacts created by the preservice teachers served as key sources of data for the research (i.e., wonder lists, inquiry units and lesson plans), as well as surveys, observations of classroom teaching, and in-depth interviews conducted with eight participants. The interviews were performed soon after graduation, and probed respondents' feelings, reactions to and perceptions of their experiences. Follow-up interviews were also conducted with three graduates half-way through their first year of teaching.

A key finding of the study was that the development of an emotive connection to science enabled the preservice teachers to envision new possibilities for learning science that were different from the ones they had experienced themselves in the past. Preservice teachers also identified several challenges to wonder in classrooms, including procedural approaches to science inquiry, a lack of time (and the idea that wonder is 'inefficient' in the context of a crowded curriculum), and pressure to perform on standardized tests. Despite these challenges, beginning teachers reported that they introduced topics through wonder and investigation in the inquiry-based units that they taught in schools. In other words, the use of wonder as a pedagogical tool during their teacher education program impacted positively on their classroom practice after graduation and developed their confidence to enact inquiry in their classrooms or adopt leadership roles in science teaching at their schools.

The authors discuss several implications for practice and further research, including the explicit use of wonder to initiate authentic disciplinary engagement in science (particularly for 'science resistant' preservice teachers); as a means of supporting more 'ambitious approaches' to science education that interrupt traditional approaches to teaching science; and as a way of developing conceptual understanding by connecting the emotive and the personal to science content. Gilbert and Byers (2017) assert that this supports a view of knowledge that is more closely aligned with the epistemic practices of scientists. The authors also invite researchers to evaluate the 'overlap' between wonder and the nature of science (NOS) in future research concerning wonder in school contexts, noting that the inclusion of wonder could enhance the conceptualization and development of more sophisticated understandings of NOS in primary contexts.

Like Gilbert and Byers (2017), Bellocchi et al.'s (2018a) study also presents a useful opportunity to examine how preservice teachers' experiences of science inquiry, in their initial teacher education, impacts their subsequent classroom practice. Bellocchi and colleagues employ narrative accounts to examine critically their own university teaching practices with secondary preservice science teachers in Australia, focusing on the flow of emotional energy arising from the enactment of science demonstrations. Drawing upon sociological theories of emotion (i.e., Collins' [2004] interaction ritual theory and a theory of eventful learning; cf. Ritchie & Newlands, 2017), each author presents a case study of their practice in which science demonstrations (demos) are viewed as first order rituals, and analyze the emotional energy that ensues across time and space when preservice teachers enact their own demonstrations—second-order rituals—in their own science teaching.

Bellocchi et al. (2018a) construct science demonstrations as effective teaching practices and sacred interaction rituals (cf. Durkheim, 1915/1964). They posit that mundane emotional energy, or the subtle emotions produced within these interaction rituals, are just as significant for understanding how preservice teachers learn to be science teachers as are more overt emotions, such as joy or excitement (cf. Davis, 2017). In this context, respect—the emotion associated with mundane emotional energy (Durkheim, 1915/1964)—compels individuals in social encounters to act in ways that are morally proper, in response to a feeling of internal moral pressure.

Bellocchi et al. (2018a) examine how the enactment of science demonstrations at university could lead to the reproduction of science demonstrations with high school students. Their analysis focuses on the experiences of a preservice teacher (Wade, pseudonym) who was allocated a teaching placement in the same school in which the third author (King) was conducting a school-based emotions study. Wade taught an 8th grade science class about the concept of air pressure by way of an egg-in-the-bottle demonstration following his first-hand experience of science demonstrations in the researchers' university classes, and the associated predict- observe-explain (POE) teaching practice that he had learnt. Bellocchi et al. (2018a) also note that, like Wade, the enactment of science demonstrations in the classes of the first and third authors (Bellocchi and King) were a second order ritual, given that the authors had learnt about them in their own initial teacher education course. In this way, the theoretical generalizability of the study is enhanced through the analysis of data from multiple teachers with different backgrounds.

Through a micro-situational analysis of a science demonstration in Bellocchi's class, in which Wade assisted, the authors demonstrate how subtle emotionality, or mundane emotional energy, is evidenced by Wade's willing participation in and anticipation of the demonstrations during the semester. Bellocchi and Wade form strong social bonds and become closely tied in their belief systems about effective science teaching, as they participate in shared symbols and practices involving science demonstrations.

The researchers also illustrate how a comment made by Wade about waiting 3-weeks for another science demo in Bellocchi's class may be viewed as a negative sanction, or reprimand, for a transgression against a valued sacred ritual. They assert that Wade's comments are suggestive of his belief that demonstrations constitute effective science teaching practices. Further, his actions, particularly through his regular voluntary participation in multiple demonstrations over time, signal respect through upholding sacred symbols and ritual practices in class.

Micro-analysis of this shared experience between Bellocchi and Wade is complemented by an analysis of Wade's classroom teaching. Analysis of three classroom episodes when Wade implemented a demonstration during placement, signal how a sequence of interactions during the egg-in-the-bottle science demonstration mediated changes from mundane emotional energy (which established students' commitment to and engagement with the activity) to more intense emotions (happiness/joy) at the culmination of the demonstration.

The authors also draw on the analysis of another case study involving Davis (second author of that study) when he was an inservice teacher. Davis presented a science demonstration showing the effect of a salinity gradient on the movement of water to his 10th grade science class. Davis had graduated from Bellocchi and King's preservice science course. Davis' demonstration was characterized by mundane emotional energy that culminated in the expression of dramatic emotion and solidarity. Davis describes how the success of his demonstration was made possible by the emotional transference from first to second order rituals; that is, his own experiences of successful demonstrations at university motivated him to replicate these experiences with his own school students. Drawing on these analyses, and those in a previous study (Bellocchi et al., 2014), the authors suggest that preservice teachers' experiences of science demonstrations as successful first-order rituals at university may empower them to reproduce demonstrations as second-order rituals in their own classes with a view to re-generate pleasant emotional climate. This reproduction of emotions and social practices in this way are important for understanding how preservice teachers learn to become effective teachers of science.

The authors conclude by posing that third-order rituals (i.e., reflective rehearsals performed introspectively in physical isolation from others) experienced by preservice teachers learning online present fertile ground for empirical studies of interaction rituals in science education; however, they note the methodological challenges of accessing research participants' first-person experiences of such rituals, given the need for participants to be phenomenologically aware.

Participation in argumentation is the focus of a Turkish study by Bahcivan (2019) who examines the relationships among preservice science teachers' scientific epistemological beliefs, epistemic emotions, and argumentativeness. Here, argumentativeness is defined as an individual's tendency to participate in argumentation. Noting that epistemological beliefs can activate epistemic emotions and subsequently predict variables such as learning strategies and learning outcomes, Bahcivan focuses on the relationship between argumentativeness and epistemological beliefs, given that few studies have explored this connection. Briefly, epistemic emotions refer to emotions that are elicited when learning processes are focused on an epistemic result, such as knowledge generation (Muis et al., 2015; Pekrun, 2006). Bahcivan posits that preservice teachers' epistemological beliefs could deeply affect their tendances to participate in argumentation, by activating their epistemic emotions.

Epistemic emotions can be activated in instances where learners experience cognitive congruence or incongruence. For example, cognitive congruence between a new learning task or knowledge can positively predict positive epistemic emotions such interest and curiosity, and negatively predict negative epistemic emotions such as boredom, confusion, and anxiety (cf. Muis et al., 2015; Trevors et al., 2017). This is significant because epistemological beliefs can activate epistemic emotions, which, in turn, influences individuals' teaching/learning conceptions.

An associational research design was employed, involving 612 Turkish preservice teachers at different stages of their middle school science teaching degrees. The instrument employed three different scales:

- scientific epistemological beliefs (26 items across 4 dimensions—certainty, source, development, and justification);
- epistemic emotions scale (13 items across 2 dimensions—pleasant emotions [curious, interested, inquisitive, amazed, and excited] and unpleasant emotions [bored, confused, anxious, frustrated, worried, muddled, monotonous and nervous]); and
- argumentativeness scale (16 items across 2 dimensions—argument approach and argument avoidance).

The epistemic emotions scale was administered to participants immediately after reading an epistemic scenario that included counter claims about nutrition and supporting scientific evidence. The aim of the scenario was to trigger epistemic activation, and the activity took approximately 25 min to complete. Data were analyzed quantitatively by way of exploratory and confirmatory factor analyses, and structural equation modelling.

The results show that preservice teachers' scientific epistemological beliefs significantly predicted their epistemic emotions. More specifically, sophisticated epistemological beliefs in the certainty and justification dimensions positively predicted their positive epistemic emotions and negatively predicted their negative epistemic emotions. Sophisticated epistemological beliefs also related positively to positive epistemic emotions in the development dimension.

The results also show that the preservice teachers' epistemic emotions significantly predicted their argumentativeness. Positive epistemic emotions aroused after reading the epistemic scenario positively predicted their argument approach and negatively predicted argument avoidance, while negative epistemic emotions related positively to argument avoidance. This finding suggests that positive epistemic emotions may motivate preservice teachers to participate in argumentation, whereas unpleasant emotions may discourage participation.

The findings of this study offer empirical evidence for the relationship between preservice teachers' emotions and their cognitions. Bahcivan (2019) argues that preservice and inservice teacher education programs aimed at developing argumentation skills must attend to epistemic emotions. This study makes an important connection to the science inquiry theme of this systemic review, given the role of argumentation in scientific inquiry (e.g., see Falk & Brodsky, 2013). Bahcivan (2019) also calls for further qualitative studies that explore the possible reasons for the relationships identified in their study. Despite the significant relationships identified between epistemological beliefs and epistemic emotions, Bahvican does not offer a critical discussion of this finding or its implications.

Five of the reviewed studies focus on inservice teachers' emotions as they engage in science inquiry in professional development programs (Davidson et al., 2020; Dreon & McDonald, 2012; Jaber et al. 2021) or in their classrooms (Cook & Buck, 2017; Ritchie et al., 2013). Like Ritchie and Newlands (2017), these studies generally report on teachers' experiences of unpleasant emotions such as anxiety, worry, frustration, fear, and anger when they initially engage in science inquiry, and how these unpleasant emotions are transformed into pleasant emotions following perseverance and the experience of success. For example, a phenomenological study by Dreon and McDonald (2012) examines the experiences of two beginning chemistry teachers in the U.S. who participated in professional development delivered by a group of practicing teachers who study science inquiry pedagogy in classrooms (the '*Invisible College for Inquiry Science Study*'). During professional development, participants analyzed their own and other teachers' practices.

Anxiety was the most salient emotion that impacted the beginning teachers' implementation of science inquiry. The teachers reported feeling anxious when they did not have deep conceptual knowledge of the content they were teaching. Conversely, they were more likely to enact inquiry pedagogy when they felt comfortable with the content. Participants also reported feeling anxiety due to the unpredictable nature of inquiry lessons, and from feeling judged by students who were frustrated by inquiry lessons. In these instances, they were more likely to adopt explicit teaching and resisted science inquiry in future classes.

Although the authors acknowledge the difference between science inquiry (i.e., the design and conduct of scientific investigations) and inquiry pedagogies (as alternatives to direct instruction), the study concerns inquiry pedagogy rather than science inquiry. A general lack of specificity about the form of inquiry instruction adopted by the two beginning teachers makes interpretation of the results difficult. At times, teachers' self-reports suggest that insufficient planning had been undertaken and their attempts at open-ended science inquiry fell flat, leading to unpleasant emotional experiences. These experiences are attributed to the inherent nature of inquiry instruction, but it is not possible, based on presented evidence, to ascertain the extent to which this was the case. Nonetheless, when teachers perceived frustration in their students due to the openness of the task, this interpretation of student emotions tended to inform changes to their instructional decisions. It is noteworthy that the openness of the task was understood as a lack of teacher explanation or direction on how to proceed with the inquiry, rather than an investigation with uncertain methods or outcomes.

Open inquiry also features as the context for Adler et al.'s (2019) study that investigates the affordances of metacognitive prompts for guiding teachers to reflect on the dynamic characteristics of open inquiry. In their study of 22 science and mathematics teachers participating in a one-year professional development course focused on teaching by inquiry, the authors report that metacognitive prompts elicit happiness, worry, surprise, frustration, satisfaction, and despair. These emotions are associated with cognitive categories, including planning, monitoring, debugging, and evaluation. For example, happiness is associated with reaching a conclusion in the planning stage of the inquiry. Worry and surprise relate to concerns about inquiry results, such as when hypotheses are not supported empirically. Performing the inquiry elicits frustration at the inquiry process and satisfaction when issues were resolved.

Alarcón-Orozco, et al. (2025) exposed early childhood preservice science teachers to an inquiry-based science education program. A three-phase instructional design was used where in phase 1, PSSTs were taught explicitly about the stages of scientific investigation (e.g., formulating a problem, designing experiment, collecting and analyzing data). In the second phase, the preservice teachers completed a scientific investigation about yoghurt production, adopting the role of school science students. Then, in phase 3, the preservice teachers reflected on their experience with the scientific investigation, before designing their own investigation aimed at children aged 3 to 6 years. The PSST self-reported their emotions using a checklist of nine pre-determined emotion labels (interest, concentration, confidence, enjoyment, insecurity, dissatisfaction, boredom, rejection, & shame), at three stages including before the scientific investigation, after it, and lastly after the scientific investigation lesson-design activity.

Alarcón-Orozco, et al. (2025) reported which of the nine emotion labels were most and least frequently associated with the anticipation of completing a scientific investigation (i.e., phase 1), enacted roles of school student and PSST (i.e., comparison of phase 2 and 3), and what stages of inquiry different emotions were associated with across phases 2 and 3. In order of frequencies reported, interest, insecurity, and concentration were anticipatory emotions in phase 1. None reported rejection, and few reported boredom, shame, or dissatisfaction. Across phases 2 and 3, as PSST shifted from school student role to PSST roles, interest, enjoyment, concentration, confidence, and insecurity were most frequent. With respect to phases of inquiry, different frequencies of specific emotion labels were reported in the student role compared to the PSST role. Interest was most common across data collection stages, and more common than enjoyment when enacting school student role. When enacting the PSST role, these two emotions and concentration and confidence appeared in similar frequencies, but none of the values were statistically significant. Overall, with stages of inquiry, interest was strongly associated with problem formulation, design of experiments, and data analysis. Problem formulation, experiment design, and variable choices were associated with insecurity. Whereas concentration was associated with choosing variables and confidence was associated with drawing conclusions.

Melville and Bartley's (2013) narrative inquiry draws on semi-structured interviews with three inservice science teachers each of whom served as a subsequent mentor of the other during a period of curriculum reform in Canada. Through the poststructuralist perspective adopted, identities are constituted when people create meaning from experiences; a process considered to be inherently emotional. Each teacher began their career by engaging in the dominant discourse of direct instruction, which was followed by adoption of inquiry teaching and learning practices. In this process of change, emotions are constructed by Melville and Bartley as being associated with *power*, a relationship that affects identity. Power to implement inquiry practices came from curriculum reform, seniority in the department, acknowledgements from other staff, and supportive colleagues who provided confidence for teachers to change practices. The need for change in practices came from emotions of concern for students and their learning for the first teacher. Seeing this concern during mentoring interactions encouraged the second teacher to change their practices. In contrast, the third teacher drew pleasant emotions from watching her students' enjoyment of inquiry tasks, whilst working in a department that had become chaired by the second teacher.

Two studies, Davidson et al. (2020) and Jaber et al. (2021), highlight the role of epistemic empathy in helping teachers to better support their own students when engaging in science inquiry. Davidson et al. (2020) investigated epistemic affect elicited by elementary teachers' experiences of engaging in scientific research, and the importance of epistemic empathy in supporting students' disciplinary engagement in science. The authors define epistemic affect as "learning to feel as scientists do when engaged in their work," arguing that teachers must have opportunities to "encounter, understand and appreciate epistemic affect in science" (p. 1008) so that they may support their students to navigate affective experiences in the classroom.

In this multi-case study, four elementary teachers participated in a 6-week *Research Experience for Teachers* (RET) program conducted at universities and national laboratories where K-12 teachers conduct scientific research through immersive 'should-to-shoulder' work with practicing scientists in a national laboratory. Using field observations and semi-structured interviews, the study reveals how participating teachers experienced and reflected on their epistemic affect. Teachers were paired with a scientist mentor and participated in laboratory group research-related activities in the morning and early afternoons, for 5-days per week. During the program, teachers actively participate in the scientists' labs, and are assigned specific roles and responsibilities (e.g., preparing samples for testing, collecting data, and co-authoring lab reports). The RET program also engaged teachers in regular cohort meetings that focus on pedagogy and supported them to reflect on their research experience with their cohort peers. The program culminates in a research poster presentation.

Three themes account for findings which revealed that the RET program afforded rich opportunities to experience epistemic affect first-hand during scientific research. First, feelings of struggle and anxiety were often experienced to begin with, as teachers sought to understand key concepts and background needed to engage in their science projects and their roles and responsibilities as part of a research team. A willingness to persevere through setbacks and uncertainty was observed during the middle weeks of the RET program, as the teachers became more adept at the procedural and analytical aspects of their research, and learned to navigate frustration due to setbacks and uncertainties. During the final phase of the RET, pride was observed as the teachers successfully communicated their understanding of their research by way of their poster presentation. Teachers also felt accomplished, knowing that their work contributed to the laboratory goals.

Davidson et al. (2020) argue that inservice teachers' affective experiences reflected the emotions experienced by scientists during their research and these experiences hold important implications for science teaching. For example, teachers should recognize, appreciate and respond to their own students' anxiety and frustration during science inquiry, and support them to "navigate their epistemic affect… potentially coming to embrace uncertainty as an exciting springboard for new inquiries and to frame intellectual challenges as interesting puzzles to solve" (p. 1035). The authors also assert that learning to normalize unpleasant emotions through firsthand experience of scientific research may support teachers to be more empathic towards their own students and support their disciplinary engagement.

Jaber et al. (2021) investigate epistemic empathy as a driver of responsive instructional practices by engaging inservice teachers as learners in the science classroom. They argue that "transforming science classrooms into communities with epistemic agency … requires a shift in teachers' practices, from proving and assessing knowledge to facilitating and orchestrating students' epistemic work" (p. 225). The context for their study was a three-phase blended science professional development program for eight upper elementary and middle school teachers, in which they collected videos from their classes for analysis. Over the course of the year-long program, the science teachers focused on their own engagement in scientific inquiry in the first phase, before studying video examples of their school students' inquiry, in the second phase. In the final phase, the teachers analyzed their students' thinking, as evidenced by the videos.

Jaber et al. (2021) found that a 'common thread' across all teachers' interviews and reflections was what they defined as "epistemic empathy" (p. 233); that is, that the teachers 'tapped into' their own experiences of 'doing science' (i.e., engaging in extended inquiry) to connect to their own students' epistemic experience. For example, the teachers reflected on their own experiences of feeling excitement in relation to an experimental result, or feeling frustrated when they were stuck, and, in a similar vein to Davidson et al. (2020), drew on these experiences to understand how their students were feeling during their own inquiries, with a view to normalize these emotions as a necessary part of doing science, and to inform how they could effectively respond. In this way, the teachers learned to attend and respond to their students' thinking as an outcome of their intellectual and emotional experiences as learners.

In Ritchie et al.'s (2013) study, the focus shifts from teachers' experiences of science inquiry in professional development to their experiences of teaching inquiry in their classrooms. Their multi-case study explores the affective experiences of four beginning science teachers in Australia who facilitate open-ended extended scientific investigations in their own classrooms. The authors explore the extent to which beginning science teachers' expectations of implementing inquiry-based activities were realized, and how their emotional states elicited by their classroom experiences mediated teaching practices. Participants reported an atmosphere of fear in their science staffrooms from their senior colleagues towards science investigations, which led them to prepare well for their own teaching. As inservice teachers' experiences with their own classes unfolded, they came to see benefits of open inquiry and found ways to scaffold their students' learning needs by conducting preliminary activities, emphasizing the importance of writing skills to support report writing, and by monitoring students' skills during the investigations. Through this scaffolding, teachers' emotions ranged from mild forms of fear, frustration, and anger. Over time, as students' confidence, skills, and independence improved, highly pleasant emotions were reported. Pleasant emotional experiences from positive student outcomes provide reassurance to these science teachers that pursuing open inquiry in their classrooms is a worthwhile pursuit.

In a study that examines the connection between science inquiry, place-based pedagogy and emotional energy, Cook and Buck (2017) draw upon sociological perspectives of emotion to explore how place-based science inquiry can foster emotional energy and solidarity for teachers and their students. Their study is underpinned by the notion that place-based pedagogy can cultivate an emotional connection to students' science learning experience.

This U.S. study involved two elementary and two middle school teachers from two different schools, who were committed to enacting place-based pedagogy in their classes. Employing a phenomenological case study design, multiple sources of data were collected by way of teacher interviews that probed their experiences implementing a place-based unit of work (pre-, mid- and post-instruction); focus group interviews with students post-instruction; audio- and video-tapes of instruction and teacher professional development sessions; the teachers' written unit plans and classroom artifacts; and the researchers' field notes. The place- based science inquiry units focused on local community problems, such as the development of an alternative energy plan to bring down the cost of power supply; evaluating the water quality of a local lake used for recreational purposes; and the restoration of native plant and animal species on the school grounds, focusing on factors such as soil quality and habitat. Collins' (2004) interaction ritual theory was used to guide coding of the data in relation to emotional energy. The researchers specifically searched for evidence of group solidarity and mutually focused attention that represented whole-class engagement with the inquiry.

Although the authors claim that teachers' (and students) emotional energy enhanced feelings of connectedness with each other and a desire to do science, there is a lack of empirical evidence for this claim, particularly given the study focuses on the school student experience and pedagogical affordances of place-based science inquiry, rather than the emotional experiences of the teachers involved. One reference is made to the pleasant emotions expressed by teachers at interview, which concerns their students' enthusiasm for their science projects. The authors claim that "communal exploration and resulting whole class discussions invoked a mood of wanting to be involved in the science investigation as evidenced by teacher reflections during and after the learning experience" (p. 153). Evidence to support this claim is not provided.

The authors refer broadly to emotional energy and a collective mood; however, they do not consider the valance of the emotional energy observed in their study (i.e., positive or negative). This is potentially significant in view of one of the study's key findings. The authors note that an important barrier to realizing the full potential of place-based learning was a lack of response from communities themselves when teachers attempted to initiate community involvement. This led to a lack of opportunities for students to interpret their inquiry data by engaging in dialogue about potential solutions to problems with key stakeholders. This could have potentially elicited unpleasant emotions from teachers, which, in turn, may have negatively impacted the emotional energy of the class.

The researchers conclude that while place-based science inquiry presents significant potential to develop emotional energy, they assert that teacher education programs should afford preservice teachers' opportunities to develop connections with local stakeholders who may be interested in place-based learning, with a view to helping overcome this barrier. No implications for preservice or inservice teacher education that concerns the emotional experiences of teachers planning or enacting place-based science inquiry are offered.

5.3 Conclusion

In this chapter, we reviewed theoretical trends and empirical outcomes of science teacher emotions research included in the theme of *science inquiry, and laboratory and practical work*. Theoretical trends range from atheoretical studies to those adopting emotion theory. Atheoretical studies ranged from those adopting lay language emotion categories, to those discussing diverse theories or emotion concepts, where authors sometimes relied on one conceptualization for emotion categories used in data collection. Theoretical studies adopted emotion theories from disciplines including psychology, biology, sociology, and postmodern perspectives.

Empirical outcomes of included studies addressed two main topics: aversion emotions in Biology practical work; and emotions associated with conducting or supporting others to conduct science inquiry. Reviewed studies provide unambiguous evidence that pleasant and unpleasant emotions are part-and-parcel of doing practical and laboratory work and inquiry in science. Although the uncertainty associated with engaging in open-ended science investigations (either through professional development or in classroom teaching) can elicit unpleasant emotions, their resolution can support preservice and inservice science teachers to develop positive views of science inquiry. Interestingly, preservice teachers' exposure to activities that elicit aversion emotions like fear and disgust, such as dissections (Fančovičová et al., 2013), as well as activities that elicit highly pleasant emotions, such as science demonstrations (Bellocchi et al., 2018a), increase their intentions to utilize these activities in their own classroom practice. Resolution of unpleasant emotions can also arise through careful planning and preparation before enacting science inquiry with school students (cf. Ritchie et al., 2013). Intentional and carefully structured science inquiry learning experiences that expose early childhood teachers to science inquiry in student roles before designing inquiries for children also helps to shift PSSTs from unpleasant to pleasant emotions (Alarcón-Orozco, et al., 2025).

Professional development involving design and implementation of science inquiry (e.g., Adler et al., 2019) can become a testing ground not only for pedagogy but also for dealing with the uncertainty associated with implementing science inquiry with school students (cf. Dreon & McDonald, 2012). Teachers' participation in science inquiry can also develop their epistemic empathy such that they can more effectively support their own students' engagement in inquiry in ways that are emotionally attuned (Davidson et al., 2020; Jaber et al., 2021). Metacognitive prompts during professional learning experiences offer one way of scaffolding science teachers' reflections on science inquiry processes and associated emotions. Mentoring and positive school student experiences with science inquiry are sources of pleasant emotions that empower teachers to use inquiry in their instruction (Melville & Bartley, 2013).

Although careful teacher preparation and prior exposure and experience with conducting scientific investigations offer key approaches to resolving teachers' unpleasant emotions that arise from their first encounters with student investigations, few studies make empirical links between preservice and inservice teachers' experiences of science inquiry, their emotions, and their intended or enacted practices in relation to teaching science inquiry in schools (cf. Bahcivan, 2019; Bellocchi, 2019; Gilbert & Byers, 2017).

References

NB Articles included in the systematic review are indicated using an asterisk (*)

*Adler, I., Zion, M., & Rimerman-Shmueli, E. (2019). Fostering teachers' reflections on the dynamic characteristics of open inquiry through metacognitive prompts. *Journal of Science Teacher Education, 30*, 763–787. https://doi.org/10.1080/1046560X.2019.1627060

*Alarcón-Orozco, M. M., Franco-Mariscal, A. J., Oliva, J. M., & Blanco-López, Á. (2025). Emotions experienced by preservice early childhood teachers during a training program in inquiry-based science education. *Journal of Research in Science Teaching*. https://doi.org/10.1002/tea.70007

*Bahcivan, E. (2019). Examining the structural relations among PSTs' scientific epistemological beliefs, epistemic emotions and argumentativeness: Sample from Turkey. *International Journal of Education in Mathematics, Science and Technology, 7*(3), 271–280.

Bellocchi, A. (2019). Emotions and teacher education. In G. Noblit (Ed.), *The Oxford research encyclopaedia of education*. Oxford University Press. https://doi.org/10.1093/acrefore/9780190264093.013.773

*Bellocchi, A., Davis, J., & King, D. T. (2018a). Science demonstrations as mediators of interaction ritual chains: Temporal transitions from mundane to intense emotional energy. In S. M. Ritchie & K. G. Tobin (Eds.), *Eventful learning: Learner emotions* (pp. 57–86). Brill. https://doi.org/10.1163/9789004377912_004

*Bellocchi, A., Ritchie, S. M., Tobin, K., King, D., Sandhu, M., & Henderson, S. (2014). Emotional climate and high quality learning experiences in science teacher education. *Journal of Research in Science Teaching, 51*, 1301–1325. https://doi.org/10.1002/tea.21170

Collins, R. (2004). *Interaction ritual chains*. Princeton University Press.

*Cook, K., & Buck, G. (2017). Our neighbourhood: A place for heightening emotional energy in science education. In A. Bellocchi, C. Quigley, & K. Otrel-Cass (Eds.), *Exploring emotions,*

aesthetics and wellbeing in science education research (pp. 141–164). Springer. https://doi.org/10.1007/978-3-319-43353-0_8
*Davidson, S. G., Jaber, L. Z., & Southerland, S. A. (2020). Emotions in the doing of science: Exploring epistemic affect in elementary teachers' science research experiences. *Science Education, 104*, 1008–1040. https://doi.org/10.1002/sce.21596
Davis, J. P. (2017). Emotions, social beings & ethnomethods: Understanding analogical reasoning in everyday science classrooms. In A. Bellocchi, K. Otrel-Cass, & C. Quigley (Eds.), *Exploring emotions, aesthetics and well-being in science education research* (pp. 121–140). Springer. https://doi.org/10.1007/978-3-319-43353-0_7
Dewey, J. (1910). Science as subject-matter and as method. *Science, 31*(787), 121–127.
*Dreon, O., & McDonald, S. (2012). Being in the hot spot: A phenomenological study of two beginning teachers' experiences enacting inquiry science pedagogy. *Teachers & Teaching: Theory and Practice, 18*(3), 297–313.
Durkheim, E. (1915/1965). *The elementary forms of the religious life* (Trans. J. W. Swain). Allen & Unwin.
Falk, A., & Brodsky, L. (2013). Scientific argumentation as a foundation for the design of inquiry-based science instruction. *Journal of Mathematics and Science: Collaborative Explorations, 13*(1), 27–55.
*Fančovičová, J., Prokop, P., & Lešková, A. (2013). Perceived disgust and personal experiences are associated with acceptance of dissections in schools. *Eurasia Journal of Mathematics, Science and Technology Education, 9*(3), 311–318. https://doi.org/10.12973/eurasia.2013.938a
*García-Ruiz, C., Lupión-Cobos, T., & Blanco-López, Á. (2022). Effects of an inquiry-based science education training program on pre-service teachers. A mixed-methods case study. *Eurasia Journal of Mathematics, Science and Technology Education, 18*(12), em2186. https://doi.org/10.29333/ejmste/12578
*Gilbert, A., & Byers, C. C. (2017). Wonder as a tool to engage preservice elementary teachers in science learning and teaching. *Science Education, 101*, 907–928. https://doi.org/10.1002/sce.21300
*Jaber, L., Dini, V., & Hammer, D. (2021). "Well that's how the kids feel!"—Epistemic empathy as a driver of responsive teaching. *Journal of Research in Science Teaching, 59*, 223–251. https://doi.org/10.1002/tea.21726
*López-Banet, L., Aguilera, D., Jiminéz-Liso, M. R., & Perales-Palacios, F. J. (2021). Emotional and cognitive preservice science teachers' engagement while living a model-based inquiry science technology engineering mathematics sequence about acid-base. *Frontiers in Psychology, 12*, Article 719648. https://doi.org/10.3389/fpsyg.2021.719648
*Melville, W., & Bartley, A. (2013). Constituting identities that challenge the contemporary discourse: Power, discourse, experience, and emotion. *Science Education, 97*, 171–190. https://doi.org/10.1002/sce.21047
Muis, K. R., Pekrun, R., Sinatra, G. M., Azevedo, R., Trevors, G., Meier, E., & Heddy, B. C. (2015). The curious case of climate change: Testing a theoretical model of epistemic beliefs, epistemic emotions, and complex learning. *Learning and Instruction, 39*, 168–183.
Pekrun, R. (2006). The control-value theory of achievement emotions: Assumptions, corollaries, and implications for educational research and practice. *Educational Psychology Review, 18*(4), 315–341.
Pekrun, R., Goetz, T., Titz, W., & Perry, R. P. (2002). Academic emotions in students' self-regulated learning and achievement: A program of qualitative and quantitative research. *Educational Psychologist, 37*, 91–105.
Pintrich, P. R., Marx, R. W., & Boyle, R. A. (1993). Beyond cold conceptual change: The role of motivational beliefs and classroom contextual factors in the process of conceptual change. *European Journal of International Relations, 63*, 167–199.
Randler, C., Ilg, A., & Kern, J. (2005). Cognitive and emotional evaluation of an amphibian conservation program for elementary school students. *Journal of Environmental Education, 37*(1), 43–52.

*Ritchie, S. M., & Newlands, J. (2017). Emotional events in learning science. In A. Bellocchi, C. Quigley, & K. Otrel-Cass (Eds.), *Exploring emotions, aesthetics and wellbeing in science education research* (pp. 107–121). Springer. https://doi.org/10.1007/978-3-319-43353-0_6

*Ritchie, S. M., Tobin, K., Sandhu, M., Sandhu, S., Henderson, S., & Roth, W.-M. (2013). Emotional arousal of beginning physics teachers during extended experimental investigations. *Journal of Research in Science Teaching, 50*, 137–161. https://doi.org/10.1002/tea.21060

Trevors, G. J., Muis, K. R., Pekrun, R., Sinatra, G. M., & Muijselaar, M. M. (2017). Exploring the relations between epistemic beliefs, emotions, and learning from texts. *Contemporary Educational Psychology, 48*, 116–132.

*Wagler, R., & Wagler, A. (2017). Understanding how preservice teachers' fear, perceived danger and disgust affects the incorporation of arachnid information into the elementary science classroom. *International Journal of Environmental and Science Education, 12*(2), 213–231.

*Wagler, R., & Wagler, A. (2022). A randomized and controlled research study assessing the emotions and beliefs of future middle school science teachers toward terrestrial isopods. *Insects, 13*(3), 233. https://doi.org/10.3390/insects13030233

Chapter 6
Emotion and Controversial Issues in Science Instruction

Controversies in science education can arise from diverse sources, including socio-scientific issues, environmental education topics, climate change discussions, nature of science debates, and evolutionary theory. These contentious areas of curriculum evoke emotions among science teachers as they navigate complex terrain that may challenge their own beliefs or those of their students. Teaching controversial issues requires pedagogical knowledge and specific skills, as science teachers must facilitate critical thinking while managing their own emotional responses and those of their students. The 15 studies reviewed in this chapter examine the emotional dimensions of teaching controversial issues in science, with research spanning various teacher education contexts and global settings. Although these topics have historically been separated in the literature, a focus on emotions provides a valuable lens through which we can understand teachers' experiences across these contentious domains. Within this chapter, we present the theoretical perspectives adopted in controversial issues research, then examine the empirical findings across five thematic areas: Socioscientific issues (SSI), environmental education, climate change, nature of science, and evolution. Across these themes, we identify key patterns in how emotions shape teachers' approaches to controversial content, their decision-making, and ultimately their pedagogical practices.

6.1 Theories of Emotion in Controversial Issues Research

The studies examining emotions in relation to teaching controversial science issues employed a diverse range of theoretical approaches. These ranged from atheoretical studies (Cebesoy & Chang Rundgren, 2023; Ekborg, 2005; Jackson et al., 1995; Palmberg et al., 2018; Rydin et al., 2025; Sullenger et al., 2000), to those examining emotions in the context of ethics and morality (Lee et al., 2012; Tuncay et al., 2012), studies adopting emotion theories from the SSI literature (Cebesoy, 2025; Cebesoy &

A. Bellocchi et al., *Emotion Research in Science Teacher Education*, Sociocultural Explorations of Science Education 32, https://doi.org/10.1007/978-3-032-07711-0_6

Chang Rundgren, 2023; Han-Tosunoglu et al., 2025), and psychological theories of emotions (Hawley & Sinatra, 2019; Hufnagel, 2015; Lombardi & Sinatra, 2013; Topçu et al., 2011; YüCel, & Özkan, 2018). Numerous studies did not adhere to a single, comprehensive theory of emotion, but rather incorporated emotional aspects into broader frameworks about learning and reasoning about controversial issues.

Most studies were atheoretical in their treatment of emotions, discussing emotions more generally without reference to specific emotion theories. For instance, Topçu et al. (2011) examined emotive informal reasoning defined as the process of negotiating and resolving controversial dilemmas using cognitive and emotional processes. Jackson et al. (1995) employed a heuristic inquiry approach acknowledging the emotional nature of topics like evolution versus religion. Sullenger et al. (2000) explored preservice teachers' emotional reactions to nature of science debates, whereas Palmberg et al. (2018) categorized emotional views about species and biodiversity; neither these authors situating their work within specific emotion theories.

Several authors examined emotions implicitly through reasoning models and socioscientific frameworks. For example, Cebesoy and Chang Rundgren (2023) used the Science, Environment, Ethics/morality, Sociology/culture, Economy, and Policy model (SEE-SEP) as their theoretical framework, with emotions defined as part of the ethical and morality dimension. Lee et al. (2012) and Tuncay et al. (2012) explored emotions through the context of empathy and moral reasoning, noting ethical or moral decisions involved personal, emotional, and relational features. Han-Tosunoglu et al. (2025) and Cebesoy (2025) examined emotive reasoning relating to the use of emotions (e.g., sympathy and empathy) when making decisions about SSI, including COVID-19 health issues and gold mining, respectively.

A few studies adopted more explicit psychological theories of emotion. Hufnagel (2015) grounded the research in the social aspect of emotions, approaching them through a sociolinguistic lens. Hawley and Sinatra (2019) and Lombardi and Sinatra (2013) incorporated concepts from self-efficacy theory and achievement emotions theory (cf. Pekrun & Stephens, 2010) related to teaching evolution and climate change, respectively.

In sum, studies on controversial issues and teacher emotions range from atheoretical, to those with theories about other constructs that refer to or define emotions, and those that adopt specific emotion theories.

6.2 Empirical Outcomes of Emotions in Controversial Issues Research

This theme represents the fourth-largest volume of emotion studies amongst the included articles, one paper more than the next most numerous theme (i.e., *Teachers' work*). This outcome indicates that knowledge about preservice and inservice teachers' emotions are heavily skewed towards these two themes, leaving ample room for research on other topics.

Within this theme, we present studies focusing on *socio-scientific issues* (SSI, Bellocchi et al., 2013; Lee et al., 2012; Topçu et al., 2011), *evolution* (Hawley & Sinatra, 2019 Jackson et al., 1995), *climate change* (Hufnagel, 2015; Lombardi & Sinatra, 2013), *environmental education* (Ekborg, 2005; Palmberg et al., 2018; Tuncay et al., 2012; Yücel & Özkan, 2018), and the *nature of science* (Sullenger et al., 2000). This brief list of topics is not surprising given it is self-evident to teachers, and others, that the nature of these topics is likely to be associated with emotions of teachers and students. The studies range from reporting the nature of emotions involved with the research topics, to those seeking to establish correlations between preservice and inservice teachers' emotions and their approach to the controversial topic.

6.2.1 Socioscientific Issues

Research on emotions in preservice and inservice science teachers unsurprisingly follows major currents of research in science education more broadly. One of the staple areas of inquiry has been on SSI. Although this is now an advanced field of investigation, only three included studies directly focus on emotions during SSI instruction. More specifically, moral reasoning (Lee et al., 2012), informal reasoning (Topçu et al., 2011), and the interplay between SSI class debates and classroom emotional climate (Bellocchi et al., 2013). These studies were distributed across South Korea, Turkey, and Australia and all three focused on secondary preservice science teachers.

Nuclear power generation, climate change, and embryonic stem cell research form the SSI topics used in Lee et al.'s study whose focus was to investigate the extent to which preservice teachers enacted ecological worldviews, socioscientific accountability, and values of social and moral compassion. It is the latter elements which have connections to the emotions because social and moral compassion requires empathy and respect for others and other species in this context. Preservice teachers approached each of the SSI topics emotionally, in Lee et al.'s words, but rarely viewed themselves as moral agents who could participate in resolving the issues. In their analyses, Lee et al. asserted how one group of preservice teachers who discussed the use of nuclear power used emotive reasoning. Like other atheoretical studies discussed in the chapter, this everyday notion of emotion being the opposite of reason was represented. The origin of this understanding of emotion is connected to Lee et al.'s framework on moral reasoning which draws on Sadler and Zeidler's (2005) study of moral reasoning and SSIs. Sadler and Zeidler discussed emotive reasoning patterns as the ways in which college students display empathy, care, and concern for the wellbeing of others.

Also drawing on ideas about reasoning in SSIs from Sadler and Zeidler (2005), Topçu et al., (2011) reported on emotions associated with informal reasoning practices of Turkish (possibly secondary school) preservice teachers. Informal reasoning involves cognitive and affective processes that people use to resolve and negotiate issues that are ill-structured. This same set of reasoning processes informs decisions to accept or reject positions or solutions related to an SSI. Three forms of informal reasoning include the categories *rationalistic, emotive,* and *intuitive.* Topçu et al. reported paired combinations of these forms in the preservice teachers' responses to SSIs including rationalistic-emotive, rationalistic-intuitive, and emotive-intuitive. Empathy and sympathy were the emotions used by preservice teachers in shaping their decision making around SSIs, like Lee et al.'s study. An example in a cloning SSI was that preservice teachers considered the future impact of reproductive cloning on a child and the parents. Empathy with the child and parents is what informed the preservice teachers' decision making. An interesting result was that although other gene therapy SSI are resolved through the rational informal reasoning approach, human cloning relies on emotive and intuitive categories of reasoning.

Building on informal reasoning research, Han-Tosunoglu et al. (2025) examined Turkish preservice biology teachers' reasoning about COVID-19 as a real-time SSI during the pandemic's first two months. This longitudinal study revealed significant emotional shifts over time, with participants' rational decision-making and moral-based reasoning decreasing while emotive and intuitive modes increased. Participants expressed various emotions including deep concern, anxiety, hope, and passion about COVID-19. Notably, even preservice teachers with relevant scientific content knowledge about viruses faced difficulty managing the emotional and psychological aspects of the crisis. The study demonstrated that decision-making became less rational and more emotionally driven as the psychological toll of extended isolation accumulated, suggesting that emotions play an increasingly important role when individuals face prolonged real-world SSI rather than hypothetical classroom scenarios. As with Lee et al.'s study above, we again see an emotion-reason dichotomy here connected to research on SSIs.

Recent research has examined how place-based approaches to SSI instruction can enhance the development of preservice teachers' emotive reasoning. Cebesoy (2025) explored Turkish preservice science teachers' participation in a six-week gold mining SSI instruction that included site visits and multiple stakeholder perspectives, finding that participants demonstrated significant shifts from emotional detachment toward deeper empathetic engagement. Before the place-based instruction, four participants showed apathy toward environmental issues, while after completion all expressed at least passive care, with many developing more sophisticated emotional responses including compassion, guilt, and anger toward environmental injustice. Like the studies by Lee et al. (2012) and Topçu et al. (2011), participants frequently used emotive reasoning when negotiating the mining controversy, particularly showing empathy and sympathy for affected communities. However, the place-based approach appeared to move the pre-service teachers beyond the local and personal perspectives noted in earlier SSI research, with teachers developing broader concerns for local environmental justice and ecological sustainability.

6.2.2 *Environmental Education*

Studies reporting emotional outcomes related to environmental education focus on constructs including elementary and secondary preservice teachers' *perceptions* (Yücel & Özkan, 2018), *reasoning* (Ekborg, 2005; Tuncay et al., 2012), and *views* (Palmberg et al., 2018). Of the four included studies, only Tuncay et al. drew on theory which recognized the role of emotions. In their study, reference was made to the affective domain from Bloom's taxonomy, defined as one's emotional response to a task. Reference was also made to scholarship that connects affect with moral reasoning. Despite this foregrounding of emotion in connection with the way preservice teachers may respond to some tasks, Tuncay et al., like all researchers in this theme, did not make explicit theoretical interpretations of their data through theory, and a specific emotion theory was not present. In this way, the trend seen in other studies of controversial issues, and other themes in this chapter, whereby researchers adopt lay perspectives on emotion is strongest in this environmental education theme. We consider this to be a major limitation because of the constraints imposed by lack of emotion theory on the nature of claims about teacher emotions in these studies. Moreover, the lay approach to emotion, as seen in other themes, tends to reproduce the emotion/reason and emotion/rationality dichotomy that has been dismissed in contemporary emotional theorizing. Having identified theoretical limitations of research in the environmental education theme, next we discuss the emotional outcomes reported by included studies.

Understanding biodiversity and sustainability depends in part on understandings about species and species identification (Palmberg et al., 2018). To understand how biodiversity and sustainability topics might be taught in schools, Palmberg et al. (2018) investigated preservice teachers views with fixed and open-ended items in a questionnaire. Results revealed that emotional views form the second most important category, by a very narrow margin, through which preservice teachers appraised the importance of species and species identification topics. Emotional views refer to emotional bonding with nature, conscious affective experiences with nature, experiences of right and wrong, or uncertain feelings. The emotional views category was dominant for preservice teachers' views on the importance of plants and animals with aesthetic aspects and humanizing animal species presenting some of the key aspects of their emotional views. Preservice teachers mainly argued about the importance of plants and animals for themselves.

Yücel and Özkan (2018) designed an intervention to shift Turkish preservice teachers' views about environmental topics. Results showed that emotions were important for the way in which preservice teachers cared for the environment. However, the intervention made no difference to preservice teachers' emotions.

As we will see later with studies focusing on climate change (e.g., Lombardi & Sinatra, 2013), some researchers are interested in the way in which teachers use scientific knowledge to reason about environmental topics (Ekborg, 2005). A local issue about the use of heat energy derived from a crematorium for district heating forms the instructional topic. Interviews with 14 elementary preservice teachers revealed that they did not react to this proposal with scientific knowledge. Ekborg (2005) writes that the preservice teachers reacted emotionally and showed keenness in reporting personal opinions. Of the atheoretical studies that reflect an emotion/reason dichotomy, Ekborg's study presented the strongest case. The lack of theory means that it was not possible for Ekborg to show how preservice teachers in her class and study were making sense of the issue emotionally. Emotion was represented as a problem, with preservice teachers being told to choose between scientific views that are pitted against ethical/religious/and emotional views in the study.

Ekborg's study evidences that preservice teachers *do* use moral reasoning when considering controversial environmental topics. Contemporary research discussed in the SSI section acknowledges the importance of emotion and moral reasoning on such topics. Tuncay et al. (2012) made moral reasoning about local and non-local environmental problems the central phenomenon in their study of Turkish preservice science teachers. Environmental dilemmas presented to preservice teachers included deforestation, e-waste, oil spills, and global warming. These topics align closely with SSI topics, however given the authors did not frame their study within the SSI field we included them in this environmental education theme.

Fifteen factors are identified by Tuncay et al. that account for preservice teachers' moral reasoning patterns. Among these were moral emotions including empathy, sympathy, respect, and conscience. Like Palmberg et al.'s findings, Tuncay et al. report that preservice teachers have an anthropocentric focus in response to environmental issues. A key recommendation made is that environmental education should consider affective and cognitive domains of moral reasoning. Moreover, Tuncay et al. recommend that environmental educators surface students' emotions regarding the value of nature rather than only focusing on transmission of ecological knowledge. This study presents a contrast to the suggestion Ekborg offers her preservice teachers about choosing between emotion or reason. Although both studies are atheoretical when it comes to emotions, the difference in perspectives is likely due to Tuncay et al.'s engagement with ideas about moral reasoning from the SSI field where the dichotomy between emotion and reason is less evident. This contrast in the two studies reinforces our claim that engagement *with* emotion theory seems to be the central feature of studies that *do not* present a dichotomy between emotion and reason/rationality.

6.2.3 Climate Change

As an emotional topic, climate change requires no introduction. Despite this, there is limited research about inservice and preservice science teachers' emotions about climate change. For instance, only three included studies directly addressed the topic (i.e., Hufnagel, 2015; Lombardi & Sinatra, 2013; Rydin et al., 2025). Although all three studies focueds on preservice teachers' emotions *about* climate change, Lombardi and Sinatra (2013) and Rydin et al. (2025) have the added focus on emotions about teaching climate change. Inservice secondary teachers and preservice elementary teachers are the participants in Lombardi and Sinatra's study. Preservice secondary science teachers participated in Rydin et al.'s study and elementary preservice teachers form Hufnagel's participants.

Emotions towards climate change and about teaching climate change were predictive of science teachers' judgements about the plausibility of climate change messages (Lombardi & Sinatra, 2013). Hopelessness had a positive relationship with plausibility and anger had a negative relationship with plausibility. The latter result suggests that unpleasant emotions reduce science teachers' evaluative judgement necessary for evaluating the evidence required for decision making about whether or not climate change is occurring.

Building on these findings about emotional influences on climate change beliefs, Rydin et al. (2025) examined Swedish preservice science teachers' beliefs about teaching climate change, revealing significant tensions around emotions in the science classroom. The study identified that preservice teachers believe climate change evokes emotions in both them and their students, with teachers experiencing deep concern, anxiety, hope, and passion about the topic. The research revealed a central pedagogical tension: preservice teachers believed climate change education should make students understand the severity of climate change while simultaneously managing unpleasant emotional impacts on students' wellbeing. The preservice teachers varied in their approaches to managing this tension, some wanted to inspire pleasant emotions through climate-friendly actions, others sought to ease unpleasant emotions by providing scientific knowledge for societal participation, whereas still others focused solely on scientific content to avoid emotional complexity. Rydin et al.'s study extends previous climate change emotion research by highlighting how the management of emotions becomes a central pedagogical challenge that preservice teachers must navigate, often without adequate frameworks for integrating emotions into their teaching practice.

The focus of Hufnagel's study is directed towards the emotional connections and disconnections preservice elementary teachers have with the topic of climate change. When preservice teachers refered to emotions in the first person, this is interpreted as a close connection with an issue, whereas use of third person means that the preservice teachers held a greater emotional distance with the topic. A hybrid approach was also reported whereby preservice teachers used the first person to describe their emotions but include hedging words that reduce the strength of connection implied by their statements. One third of expressions corresponded with impact of climate change with impacts on other species being the dominant category. This focus on impacts upon other species or future generations of humans was seen as a way in which preservice teachers othered their emotions. Fear was the emotion associated with othering. The preservice teachers also engaged emotionally with climate change as a forthcoming event rather than something that had happened or was currently happening. Emotions related to lack of action on climate change were directed at self, other, or collectives. Anger was the emotion used when blaming the human collective for climate change, which served to reduce the focus on the preservice teachers' contribution to the problem. Guilt characterized their individual role in affecting climate change but blame for climate change was mainly directed at the collective. An interesting question identified by Hufnagel was what role emotions about self and collectives had to play in the plausibility beliefs studied by Lombardi and Sinatra (2013). This is one variable that was not part of Lombardi and Sinatra's study, which could add further refinement to the connections among factors that shape preservice teachers' dispositions towards climate change and climate change instruction. One teacher in Lombardi and Sinatra's study commented on an attempt to invoke strong emotions in her students to motivate their actions on environmental issues. Understanding how teachers, teacher educators, and students use emotions in science learning and teaching presents an obvious next step for research given we know little about what people *do* with emotions and what is *done to* people's emotions by larger social structures.

6.2.4 Nature of Science

Nature of science remains a contentious topic within science education research; however, preservice teachers are seldom exposed to such debates (Sullenger et al., 2000). Beginning from this premise, Sullenger et al. (2000) reflected on their instruction by focusing on the cognitive and emotional reactions of preservice science teachers' during their first encounters with nature of science debates. Instruction focused on the culture wars; the debates from the 1960s onwards that variably challenged the authority of science and the essence of its epistemological claims. Mixed observations were reported including that some preservice teachers found the intensity of interaction around debates traumatic in one extreme, uneventful at another, and yet others welcomed the experience. In the past, related topics triggered anxieties, and the experiences with Sullenger et al.'s class provided scope for discussing those and gaining new confidence to speak about the topic. Preservice teachers with religious beliefs found comfort in arguments that challenged the authority of science, and the debates helped these students articulate their resentment and distrust towards science.

These students also met with hostile reactions from peers leaving religious students feeling dismayed. Overall, the course provided preservice teachers with scope to find their voice in relation to the debate and for many, to take up different positions in the available lines of argument.

The range of emotions constructed by Sullenger et al. point to interesting areas for future inquiry through well-theorized studies that are designed methodologically and conceptually to address the role that emotions play in nature of science debates. Experiences of students with religious beliefs and associated emotions is a matter addressed more directly in two studies, described below, reporting emotional outcomes related to teaching evolutionary theory.

6.2.5 *Evolution*

A geographically localized issue of teaching evolution in the Southern and South-Western United States is the focus of Hawley and Sinatra (2019) and Jackson et al.'s (1995) studies, whose publication years indicate the ongoing interest in this topic despite being the only two included studies sharing a focus on emotion. Teachers face emotional challenges when teaching evolution to students with orthodox religious beliefs, and teachers themselves may share these beliefs, creating personal conflicts between religious convictions and scientific content. Both studies addressed these emotional dimensions but from different perspectives, Jackson et al. presented an early exploration of emotional issues through stakeholder viewpoints, while Hawley and Sinatra reported on outcomes of a professional development intervention.

Jackson et al. (1995) explored the relationship between orthodox Christian beliefs and evolutionary theory by drawing together diverse perspectives of scientists, science educators, and prospective teachers. As an evolutionist and secular humanist, the lead author, Jackson, reflected on how he underestimated the level of emotional involvement his students had invested in their religious beliefs as they questioned evolutionary theory in discussions. Science education graduates involved in the study reported substantial changes in their views toward evolutionary theory after engaging with the course readings and interviews as part of the research, with some experiencing strong emotional reactions as they could no longer reconcile their past beliefs with their new understanding of evolution. Jackson et al. concluded that science education must be sensitive to these issues to avoid doing serious harm.

Inservice teachers in Hawley and Sinatra's (2019) professional development study expressed concerns about community reactions, being ostracized from religious and family groups, reactions of school administration, and personal epistemological conflicts about evolutionary theory. The workshop developed by Hawley and Sinatra proved effective at enhancing inservice teachers' self-efficacy and confidence, reducing unpleasant emotions about evolutionary theory and about students' questions, and increasing pleasant emotions toward teaching evolution. Their study demonstrates that respectfully crafted professional development can reduce teachers' worries about teaching evolution, which is important because teachers are more likely to teach the topic when they feel efficacious and confident and their fears are allayed.

The positive results that Hawley and Sinatra report provide confidence that emotional challenges when teaching evolution in unsupportive environments can be addressed through targeted professional development. A key development across the two studies is the move from Jackson et al.'s atheoretical approach to Hawley and Sinatra's engagement with emotion theory in education, reflecting a general trend toward more theoretically grounded emotion research since the 1990s.

6.3 Conclusion

In this chapter, we reviewed theoretical trends and empirical outcomes of science teacher emotions research included in the theme of emotion and controversial issues in science instruction. Studies included those that are atheoretical (Palmberg et al., 2018; Sullenger et al., 2000; Jackson et al., 1995; Rydin et al. 2025), to those discussing diverse theories or emotion concepts without explicit theoretical frameworks (Topçu et al., 2011; Ekborg, 2005). Theoretical studies adopted emotion theories from disciplines including psychology and education, with some drawing on self-efficacy theory and achievement emotions theory (Hawley & Sinatra, 2019; Lombardi & Sinatra, 2013), SSIs frameworks (Cebesoy, 2025; Cebesoy & Chang Rundgren, 2023; Han-Tosunoglu et al., 2025), and sociolinguistic approaches to emotion (Hufnagel, 2015).

Empirical outcomes of reviewed studies addressed five main areas: SSIs, environmental education, climate change, nature of science, and evolution. The studies provide clear evidence that controversial topics in science instruction inevitably can evoke strong emotional responses in both preservice and inservice teachers, with implications for their pedagogical decision-making and classroom practices. Several overarching patterns emerge across the controversial issues domain. Emotions play a central role in teachers' reasoning about controversial topics, with studies consistently documenting the interplay between emotional, rational, and intuitive reasoning processes (Lee et al., 2012; Topçu et al., 2011; Han-Tosunoglu et al., 2025). Empathy, compassion, and sympathy emerge as particularly significant emotions that shape teachers' decision-making across multiple controversial domains, from SSI to environmental concerns (Lee et al., 2012; Topçu et al., 2011; Tuncay et al., 2012).

The research reveals significant pedagogical tensions around emotion management in controversial science instruction. Teachers must navigate complex terrain where they simultaneously seek to engage students emotionally with important issues while managing potentially distressing emotional impacts (Rydin et al., 2025). This is particularly evident in climate change education, where teachers experience deep concern and anxiety about the topic while struggling to balance emotional engagement with student wellbeing. Place-based approaches to controversial science instruction show particular promise for developing teachers' emotive reasoning capabilities, with research demonstrating that direct engagement with local controversial sites and affected stakeholders can move teachers from emotional detachment toward more sophisticated forms of empathetic engagement (Cebesoy, 2025).

Personal beliefs and worldviews significantly influence teachers' emotional responses to controversial content, most clearly documented in evolution education, where teachers with religious beliefs experience emotional conflicts between their personal convictions and scientific content (Jackson et al., 1995). The studies suggest that targeted professional development can effectively address emotional barriers to teaching controversial content. For example, Hawley and Sinatra's (2019) evolution workshop demonstrated that respectful, informative professional development can reduce unpleasant emotions and increase teachers' self-efficacy and confidence in addressing contentious topics. However, teachers often lack adequate frameworks for integrating emotions into their teaching practices, particularly when managing the emotional dimensions of controversial issues (Rydin et al., 2025). The effectiveness of place-based approaches suggests that authentic engagement with real-world controversies may provide more powerful contexts for developing these integration skills than abstract discussions alone (Cebesoy, 2025).

The prevalence of the emotion-reason dichotomy is particularly evident in studies drawing on informal reasoning frameworks that categorize reasoning into rational, emotional, and intuitive types. It appears that contemporary understandings of emotion are yet to influence thinking in the SSI field about emotion and rationality, and this is an immediately obvious conceptual area that future studies can and should address.

References

NB Articles included in the systematic review are indicated using an asterisk (*)

*Bellocchi, A., Ritchie, S. M., Tobin, K., Sandhu, M., & Sandhu, S. (2013). Exploring emotional climate in pre-service science teacher education. *Cultural Studies in Science Education, 8*, 529–552. https://doi.org/10.1007/s11422-013-9526-3

*Cebesoy, Ü. B. (2025). Exploring preservice science teachers' emotive reasoning through place-based SSI instruction: The case of gold mining. In B. Namdar & E. Karahan (Eds.), *Socioscientific issues focused teacher education* (pp.153–180). Springer. https://doi.org/10.1007/978-3-031-55233-5_10

*Cebesoy, Ü. B., & Chang Rundgren, S. N. (2023). Embracing socioscientific issues-based teaching and decision-making in teacher professional development. *Educational Review, 75*(3), 507–534. https://doi.org/10.1080/00131911.2021.1931037

*Ekborg, M. (2005). Is heat generated from a crematorium an appropriate source for district heating? Student teachers' reasoning about a complex environmental issue. *Environmental Education Research, 11*(5), 557–573. https://doi.org/10.1080/13504620500169718

*Han-Tosunoglu, C., Özer-Aker, F., & Ağlarcı-Özdemir, O. (2025). Transformative shifts in informal reasoning of pre-service teachers in a real-time health issue. *Journal of Science Teacher Education, 36*(4), 1–35. https://doi.org/10.1080/1046560X.2024.2404339

*Hawley, P. H., & Sinatra, G. M. (2019). Declawing the dinosaurs in the science classroom: Reducing Christian teachers' anxiety and increasing their efficacy for teaching evolution. *Journal of Research in Science Teaching, 56*(4), 375–401. https://doi.org/10.1002/tea.21479

*Hufnagel, E. (2015). Preservice elementary teachers' emotional connections and disconnections to climate change in a science course. *Journal of Research in Science Teaching, 52*, 1296–1324. https://doi.org/10.1002/tea.21245

*Jackson, D. F., Doster, E. C., Meadows, L., & Wood, T. (1995). Hearts and minds in the science classroom: The education of a confirmed evolutionist. *Journal of Research in Science Teaching, 32*, 585–611. https://doi.org/10.1002/tea.3660320606

*Lee, H., Chang, H., Choi, K., Kim, S.-W., & Zeidler, D. L. (2012). Developing character and values for global citizens: Analysis of pre-service science teachers' moral reasoning on socioscientific issues. *International Journal of Science Education, 34*(6), 925–953. https://doi.org/10.1080/09500693.2011.625505

*Lombardi, D., & Sinatra, G. (2013). Emotions about teaching about human-induced climate change. *International Journal of Science Education, 35*(1), 167–191. https://doi.org/10.1080/09500693.2012.738372

*Palmberg, I., Hermans, M., Jeronen, E., Kärkkäinen, S., Persson, C., & Yli-Panula, E. (2018). Nordic student teachers' views on the importance of species and species identification. *Journal of Science Teacher Education, 29*, 397–419. https://doi.org/10.1080/1046560X.2018.1468167

Pekrun, R., & Stephens, E. J. (2010). Achievement emotions: A control-value approach. *Science and Personality Psychology Compass, 4*, 238–255. https://doi.org/10.1111/j.1751-9004.2010.00259.x

*Rydin, M., Gericke, N., Christenson, N., & Haglund, J. (2025). Areas of tension in teaching climate change–conflicts between Swedish preservice science teachers' beliefs. *International Journal of Science Education*, 1–25. https://doi.org/10.1080/09500693.2025.2466771

Sadler, T. D., & Zeidler, D. L. (2005). Patterns of informal reasoning in the context of socioscientific decision making. *Journal of Research in Science Teaching, 42*, 112–138.

*Sullenger, K., Turner, S., Caplan, H., Crummey, J., Cuming, R., Charron, C., & Corey, B. (2000). Culture wars in the classroom: Prospective teachers question science. *Journal of Research in Science Teaching, 37*, 895–915. https://doi.org/10.1002/1098-2736(200011)37:9%3c895::AID-TEA3%3e3.0.CO;2-3

*Topçu, M. S., Yilmaz-Tuzun, O., & Sadler, T. D. (2011). Turkish preservice science teachers' informal reasoning regarding socioscientific issues and the factors influencing their informal reasoning. *Journal of Science Teacher Education, 22*, 313–332. https://doi.org/10.1007/s10972-010-9221-0

*Tuncay, B., Yılmaz-Tüzün, Ö., & Teksoz, G.-T. (2012). Moral reasoning patterns and influential factors in the context of environmental problems. *Environmental Education Research, 18*(4), 485–505. https://doi.org/10.1080/13504622.2011.630576

*YüCel, E. Ö., & Özkan, M. (2018). Analysis of change in the environmental perceptions of prospective science teachers: A longitudinal study. *Asia-Pacific Forum on Science Learning and Teaching, 19*(1).

Chapter 7
Emotion and Technology in Science Teacher Education

The eight included studies that report emotional outcomes of preservice or inservice science teachers in relation to technology address web-based instruction or online learning experiences (Bellocchi et al., 2016; Eilon & Kliachko, 2004); understanding the potential to develop online communities of practice through *WhatsApp* groups that enhance teachers' professional development (Cansoy, 2017); the efficacy of different online teaching approaches, including gamification (Sanchez-Martin et al., 2018); flipped classroom models (Jeong et al., 2018); preservice teachers' experiences and perceptions of teaching children via distance education during the COVID-19 pandemic (Şevki Ayvaci et al., 2023); the relationship between emotional intelligence, learning styles, and science and technology efficacy on PSSTs' academic performance (Amponsah et al., 2024); and the effects of Technological Pedagogical Content Knowledge (TPCK), self-efficacy, enjoyment, control appraisal and value appraisal on preservice and inservice science teachers' integration of Information and Communication Technologies (ICTs) (Abualrob & Al-Saadi, 2025). Study participants ranged from high school science preservice and inservice teachers, high school science teachers, and elementary preservice teachers, with research conducted in Australia, Spain, Turkey, Israel, Palestine, and Ghana.

One study uses a single qualitative method (participatory observation of written artifacts—'virtual ethnography'; Cansoy, 2017), three use a multi-method qualitative approach (Bellocchi et al., 2016; Eilon & Kliachko, 2004; Şevki Ayvaci et al., 2023), one employs a quantitative descriptive approach using structural equation modelling (Abualrob & Al-Saadi, 2025); and three use quantitative methods involving questionnaires (Amponsah et al., 2024; Jeong et al., 2018; Sanchez-Martin et al., 2018). Amponsah et al. (2024) adopts a descriptive correlational research design to examine how PSSTs' emotional intelligence impacts their academic performance. Whereas Bellocchi et al. (2016), Cansoy, (2017), Eilon and Kliachko (2004), and Şevki Ayvaci et al. (2023) specify research designs including interpretive netnography (Özüdoğru, 2014), action research, hermeneutic phenomenological methodology and qualitative case study. Jeong et al. (2018) and Sanchez-Martin et al. (2018) (both quantitative

A. Bellocchi et al., *Emotion Research in Science Teacher Education*, Sociocultural Explorations of Science Education 32, https://doi.org/10.1007/978-3-032-07711-0_7

studies) do not specify a design and focus on applying questionnaires and quantitative analysis of the data obtained.

7.1 Theories of Emotion in Research on Technology

Bellocchi et al. (2016) adopted sociological theories of emotion to define and frame the phenomenon in their study (i.e., Turner [2007], Collin's [2004] Interaction Ritual Theory, and emotional energy [Durkheim, 1915/1965), while Amponsah et al. (2024) and Abualrob and Al-Saadi (2025) adopted psychological perspectives of emotion (i.e., control-value theory [CVT; Pekrun, 2006] and emotional intelligence [Goleman, 1995], respectively). All other studies were atheoretical. Sanchez-Martin et al. (2018) discuss different ideas in the literature about emotion drawing chiefly on emotions research by other Spanish scholars, as do Jeong et al., (2018); however, they do not articulate a defined theoretical choice and with no clear influence of the discussed ideas in the interpretation of emotions in the data. Similarly, Şevki Ayvaci et al. (2023) confine their use of the term *emotions* to feelings experienced by preservice teachers during the planning and delivery of distance education (i.e., "distance education emotions", p. 140) without defining a theoretical choice. Two studies refer to the notions of *emotional support* in the context of online learning. Cansoy (2017) investigated the way in which teachers emotionally supported one another through an online app, drawing on previous research concerned with informal online communities and networks as a source of teacher professional development. Similarly, Eilon and Kliachko (2004) found that emotional support was a key theme to emerge from their identification of the teacher's role in web-based instruction, and how these roles were perceived by elementary preservice science teachers. Bellocchi et al. (2016) also explore the notion of emotional support; however, this exploration is grounded in Turner's (2007) work on the formation of social bonds and the development of group solidarity.

Overall, the study of preservice and inservice science teachers' emotions in relation to technology would be enhanced by making clear research design choices and purposefully engaging with theories of emotions. Although it is not necessary for all research designs to involve the use of specific emotion theories for interpretation of data, those designs that do not require a priori theorizing were not evident in the included studies, and, moreover, if they had been, it would be necessary to articulate the metatheoretical approach (i.e., paradigm, worldview, perspective) in the research design chosen. For example, phenomenological, post- structuralist, and ethnomethodological studies do not require, and may eschew, the need for theory. Better articulation of research design choices is evident in the two quantitative studies, where questions of validity and reliability cannot be answered without clear design choices being specified, as is the case with the other studies.

7.2 Empirical Outcomes of Emotions Research on Technology

Each included study reports a range of outcomes and findings. Where online learning communities and web-based instruction are concerned, the use of technology in achieving emotional support features across three studies (i.e., Bellocchi et al., 2016; Cansoy, 2017; Eilon & Kliachko, 2004). The study by Bellocchi et al. (2016) investigated the formation, maintenance, and disruption of social bonds among a group of 64 secondary preservice science teachers in Australia who studied a science pedagogy course either face-to-face ($n = 16$) or online ($n = 48$). Online learning experiences included video recordings of discrepant events taught face-to-face, online course readings, video recordings of face-to-face lectures, online discussions of videos of science classroom instruction, and the development of concept maps. Learning resources are made available via an online learning management system, which includes a weekly discussion board topic to mediate online learners' interactions through instructor questions. For research purposes, a question about emotional experiences associated with online learning was included on the discussion board (i.e., an *e-motion diary*). In total, 284 blog entries were posted, containing 392 emotional reports.

Across all five weeks of data collection, the online learners used a range of labels to identify their emotional experience associated with social bonds: *anxiety, comfort, involved, confident/strong, apprehension, overwhelmed, isolation, concern, enjoyment, fear, uncertainty, excited, frustrated, unsupported, lack of fear,* and *connection.* Preservice teachers learning online reported feelings of inclusion in the face-to-face group when they watched videos of classroom transactions, as the recordings provided experiences of social contact. Feelings of isolation and alienation arose, however, from poor quality lecture and demonstration recording, and the lack of opportunity for live discussion, which served to disrupt social bonds between online students, their face-to-face peers, and the instructor. The study also found that the e-motion diaries facilitated social interactions focused on shared emotional experiences which, in turn, led to the formation of social bonds through the expression of empathy, which alleviated feelings of fear, trepidation and anxiety about becoming science teachers. For example, comments such as *"We are all in the same boat"* provided a sense of solidarity and emotional connection, and preservice teachers reported feeling less isolated because of these shared emotions.

Cansoy's (2017) study analyzed inservice science teachers' messages, called *sharings,* via the messaging app, *WhatsApp*^{tm}, with a view to glean insights into how they might contribute to one another's professional development in an online community of practice. The biology, chemistry, physics, and general science teachers ($n = 12$) taught at a private secondary school in Turkey and had a range of years of teaching experience spanning from 5 to 22 years. Content analysis of their messages ($n = 1363$), which included text, sound, photographs, and videos, revealed four themes that supported their professional development: sharing of disciplinary science knowledge, sharing of pedagogical content knowledge, sharing of in-school science teaching practices, and sharing for emotional support. The author found that, through their online community of practice, the teachers emotionally supported each other through their posts. For example, they would celebrate their teaching achievements in a bid to develop mutual trust, respect, and support. This was expressed through text and symbols (i.e., emojis). They would also plan social activities as a way of fostering solidarity and to encourage participation through their posts. Notwithstanding the opportunities for science professional development, Cansoy (2017) concluded that the app provided valuable opportunities for building social and emotional support in an online community of practice.

In contrast to peer emotional support reported by Bellocchi et al. (2016) and Cansoy (2017), Eilon and Kliachko (2004) identify the teacher educator's roles when delivering a web-based human biology course to 13 Israeli preservice teachers and their perceptions of the course. The biology course included learning resources such as a dedicated website, textbooks, videos, computer simulations, and online discussion forums. Data were collected over an 11- week period and included teacher and student emails, online discussions, and interviews. Analysis revealed that the provision of guidance, mediation, management, technology support, and emotional support were the key roles played by the teacher educator. The emotional support category represented 3% of students' emails ($n = 3$). For example, students felt that the teacher provided encouragement, consolation, or empathy when they reported difficulties learning and engaging online, or at times, disapproval. Analysis of teacher emails revealed that 6% ($n = 11$) concerned the provision of emotional support; for example, by acknowledging students' anxiety about learning online, and offering reassurance. Overall, the study found that developing a friendly web-based environment for elementary preservice teachers that nurtured social relationships, enhanced motivation, and provided encouragement to individual students were important in offering emotional support during the course.

A more recent study by Şevki Ayvaci et al. (2023) explored the emotional responses of Turkish preservice science teachers required to develop and teach a science program to elementary children via distance (online) education during the COVID-19 pandemic. Eight preservice science teachers in their final year of study (4 male, 4 female) participated in the qualitative study. Data were collected via observations and semi-structured interviews. The participants reported a range of emotions in relation to distance education, which were organized into four categories: anxiety ($n = 2$), fear ($n = 4$), curiosity ($n = 1$), and sadness ($n = 1$). Anxiety and fear were articulated in terms of feeling nervous, depressed, and afraid about teaching online due to perceptions about having inadequate technological pedagogical content knowledge. Curiosity manifested in one participant's use of a new online system, whereas another expressed sadness/unhappiness about the prospect of not being able to see the students that they would be teaching. The authors point to the need for appropriate preservice science teacher training, which prepares them to deliver distance education effectively by avoiding unpleasant emotions.

Across the four studies reviewed so far, data sources provided opportunities for participants to report their emotions openly and in unstructured ways. In Bellocchi et al.'s (2016) study, adoption of emotion theory meant that emotion labels were used to interpret participants' comments that did not specify an emotional label. For Cansoy (2017), Eilon and Kliachko (2004), and Şevki Ayvaci et al.'s (2023) studies, a more general discussion of emotion is presented, and this is most likely due to the absence of a guiding theory that could be used to interpret the data in the context of broader understandings of emotion. In contrast to these four studies, the next two studies discussed adopt questionnaires that involve choice of emotion labels from specified options in the instruments used.

Jeong et al. (2018) focused on the effects of a flipped learning approach on learning outcomes, perceptions, and emotions evident in Spanish elementary science teacher education classes over two years. 153 preservice teachers (63% male; 37% female; average age 21) studying matter and energy topics in a general science course participated in the study. Participants had different disciplinary backgrounds, classified as either a science and technology background or an arts and social sciences background. The researchers employed a post-test questionnaire to elicit participants' emotions after the flipped learning experiences and a two-sample *t*-test for data analysis. Specified emotion items were used as follows: fun, confidence, enthusiasm, and tranquility (pleasant emotions); and nervous, concerned, bored, and fearful (unpleasant emotions). Participants rated the frequency of their chosen emotion labels on a scale of 1–10 (low to high). Additional data about achievement on course content were taken from the regular classroom assessment and tasks which were summarized as an average student result at the end of the course.

The researchers found that pleasant emotions were overwhelmingly reported in relation to the flipped learning approach. Of the pleasant emotions, *fun* and *enthusiasm* are reported most frequently and *confidence* received the lowest score. Of the unpleasant emotions, *concern* was reported most frequently, with the lowest score corresponding to *boredom*. Women generally expressed higher values in relation to unpleasant emotions; however, no significant differences were observed in relation to pleasant emotions based on gender. Preservice teachers with a science and technology background reported significantly lower levels of unpleasant emotions; however, there were no significant differences between the two disciplinary groups in terms of reported pleasant emotions, which were consistently high. The authors assert that the flipped learning model had a significant impact on preservice teachers' learning outcomes and pleasant emotions towards the course, particularly when they did not have a science education background.

Although Jeong et al. (2018) report interesting results, the study does not engage with literature regarding problems associated with interpreting self-reported emotions data and does not allow for critical discussion of the over-reporting of pleasant emotions compared to unpleasant emotions in their study. Conclusions about pleasant emotions being associated with the flipped learning approach, therefore, cannot be substantiated. Future research could redress this issue through closer engagement with studies of emotions and better understandings of alternative factors that influence self-report. Similarly, the claim that preservice teachers without a science and technology background experienced the greatest emotional benefit from flipped instruction cannot be substantiated without comparison to general trends in emotional reporting of people with diverse backgrounds. Lastly, given that the questionnaire is administered at the end of the course, discussion is needed about the extent to which emotions represent global evaluations of the course or specific moments in time. Absence of an emotion theory also makes it challenging to understand how the emotion labels used in the questionnaire are obtained and to what extent they are emotions or understood to be emotions within the Spanish cultural context.

Sanchez-Martin et al. (2018) report another study in a Spanish elementary preservice science education context. Their focus is on understanding how different teaching methods affect preservice teachers' emotional performance when dealing with science. Inclusion of this study in the technology theme is based on one of the teaching methods which involves gamification. The other approaches tested were teacher lecture, teacher lecture with gamification, teacher lecture with audio visual scaffolds, guided inquiry, and practical work. Gamification consisted in using an online quiz generator (i.e., Kahoot!^{tm}). Sanchez-Martin et al.'s (2018) study draws on perspectives from positive psychology dividing emotions into either positive or negative, static, or dynamic. The underlying assumption in this approach is that learning should be a pleasant emotional experience. Based on this assumption about emotions, a questionnaire is designed and administered across 17 lessons to 120 elementary preservice science teachers studying a course about the natural environment in their final year of studies. Details of questionnaire validation are not provided. During lessons, the questionnaire involves preservice teachers holding up a card which represents their choice from one of four possible emotion labels: rejection (negative-dynamic), boredom (negative-static), satisfaction (positive-static), or surprise (positive-dynamic). Participants also provided reasons for their chosen emotions by choosing from one of the following pre-specified choices: personal disposition; teaching approach; subject taught; or other. It is not clear how these choices were made by participants or how the researchers collected them.

Three key outcomes were drawn from the analyses: (1) there were significant differences in emotions associated with the different teaching methods; (2) practical work and lectures with gamification led to experiences of surprise and lectures with audiovisual supports were boring; and (3) guided inquiry was not advised as a method by which to engage students. After the type of teaching methods used, the nature of lesson content was the next most frequent factor influencing the types of emotion label chosen by participants, with lessons involving mathematical content receiving the greatest frequency of unpleasant emotions. The authors claim that their key contribution is the method of continuous measurement of emotions during instruction; however, this claim is tempered by a lack of clarity regarding the methodological design. Moreover, the level of variability within each teaching method makes it challenging to draw conclusive claims about the relationship between emotions and the teaching methods studied.

Amponsah et al. (2024) examine how PSSTs' emotional intelligence impacts their academic performance, alongside other factors such as science and technology self-efficacy and learning styles. Emotional intelligence may be considered a non-cognitive skill that incorporates self-awareness and self-regulation of one's emotions, empathy (i.e., understanding others' emotions), as well as motivation and social skills, such as interpersonal communication (cf. Goleman, 1995). The authors assert that emotional intelligence is essential to balance the emotional and cognitive demands of teaching and supports PSSTs to perform better during their initial teacher education. The authors also draw on previous research that indicates a positive correlation between emotional intelligence and learners' academic performance (e.g., Rabi et al., 2021; Salovey & Mayer, 2020).

Preservice science teachers ($n = 91$) in the University of Ghana's School of Education completed a questionnaire comprised of three scales:

- The *STEM Self-Efficacy in Science and Technology Scale* (Wang & Degol, 2017), which measures participants' confidence in performing STEM-related tasks.
- An emotional intelligence scale, based on the *Trait Emotional Intelligence Questionnaire* (TEIQue) (Petrides et al., 2006), which comprises 13 Likert-style items that examine domains such as self-awareness, self-regulation, and empathy on a 5-point scale (1 = strongly disagree, 5 = strongly agree).
- A learning style scale, based on the *VAK Learning Styles Model* (Barbe et al., 1979), which categorizes participants into three learning styles: visual (V), auditory (A) and kinesthetic (K).

A panel of experts validated the questionnaire (emotional intelligence researchers) before it was piloted with 40 preservice teachers. Reliability was found to be satisfactory (Cronbach's $\alpha = 0.78$). Participants' academic performance was measured by way of their Grade Point Averages (GPA). Questionnaire data were analyzed using descriptive statistics (i.e., means and standard deviation) to determine the impact of the independent variables on academic performance, whereas correlation analysis and hierarchical multiple regression were used to analyze associations between variables.

Although the study was guided by several research objectives, including the examination of gender differences on participants' science and technology self-efficacy, only those relevant to emotional intelligence are reported here. Emotional intelligence ($M = 2.98$) was found to have a medium impact on academic performance, whereas science and technology self-efficacy ($M = 3.97$) had the greatest impact (particularly for male students), and learning styles ($M = 2.23$) the least impact. A slight negative correlation was found between academic performance and emotional intelligence ($r = -0.081$), but it was not significant ($p = 0.447$), which suggests that PSSTs with higher emotional intelligence have somewhat poorer academic performance. The study also found that gender did not significantly moderate emotional intelligence and academic performance.

The authors note that previous research concerning emotional intelligence and academic performance has yielded mixed results in different educational contexts, such as engineering, medical, and nursing education, as well as high school and college settings. The authors recommend that initial teacher education programs integrate opportunities to develop students' emotional intelligence skills, arguing that high emotional intelligence is still an important trait, supporting improved interpersonal interactions, greater empathy, and effective communication; however, their study did not offer any empirical evidence to support these assertions.

Collectively, Amponsah et al.'s (2024) findings illustrate the complex and context-dependent nature of non-cognitive traits in influencing PSSTs' academic performance. It is noted that the current study was the only one that examined the construct of emotional intelligence. In view of the findings, research conducted in diverse contexts with larger sample sizes is warranted to generate further empirical evidence regarding the significance of emotional intelligence for pre- and inservice science teachers' practice. Further research is required to test the authors' claim that teachers with high emotional intelligence have improved interpersonal interactions, empathy, and effective communication in the science classroom. However, given the number of available critiques of the emotional intelligence construct, it would be worthwhile for researchers to consider and address critiques before conducting further research.

Abualrob and Al-Saadi (2025) adopt a descriptive quantitative approach using SEM to examine the predictive relationships and interactions among five constructs—TPCK, ICT teaching self-efficacy, enjoyment, control appraisal, and value appraisal—evaluating differences between preservice and inservice science teachers. The authors draw on CVT to explore how emotions and attitudes can be predicted by control and value appraisals regarding the integration of ICTs in science classrooms (Pekrun, 2006). Enjoyment is a particular focus, given that teachers who experience positive feelings in relation to ICTs are more likely to integrate them into their practice (cf., Goetz et al., 2006). In this study, control appraisal is the perceived ability to use ICTs, whereas value appraisal is the perceived significance of ICT in teaching. Collectively, these appraisals determine how teachers implement ICTs in their science teaching practice (Abualrob & Al-Saadi, 2025).

Conducted in Palestine, the study employed stratified random sampling to recruit preservice and inservice teachers to complete an electronic questionnaire based on validated scales from previous studies (i.e., Mishra & Koehler, 2006; Venkateshet al., 2003). The questionnaire included multiple items structured around the five dimensions of TPCK, ICT self-efficacy, enjoyment, control appraisal, and value appraisal. Composite reliability and factor loadings demonstrated satisfactory internal consistency and construct reliability. Valid questionnaire responses were provided by 69 preservice teachers, and 172 inservice teachers (70% and 82% response rates, respectively). SEM was employed to test five hypotheses that concern the following path relationships:

TPCK, self-efficacy, enjoyment, control → Value appraisal (H1)
TPCK, self-efficacy, enjoyment, value → Control appraisal (H2)
TPCK, self-efficacy, value, control → Enjoyment (H3)
TPCK, value, enjoyment, control → Self-Efficacy (H4)
Self-efficacy, enjoyment, value, control → TPCK (H5)

A sixth hypothesis (H6) predicted a significant difference between the responses of inservice and preservice teachers regarding control appraisal, enjoyment, ICT self-efficacy, TPCK, and value appraisal.

Analysis revealed that H1-H5 were supported with statistically significant path coefficients, with ICT teaching self-efficacy, control appraisal, and enjoyment being the strongest predictors of value appraisal and successful ICT integration. In relation to H6, the only statistically significant difference between the two groups of participants was found in relation to TPCK; that is, inservice teachers had significantly higher TPCK ($M = 4.15$) than preservice teachers ($M = 3.96$), $p = 0.029$. No significant differences were found in self-efficacy, enjoyment, control, or value appraisal. This finding suggests that experience in real classroom settings contributes to stronger TPCK development, whereas other affective factors may develop similarly during training. ICT self-efficacy, TPCK, enjoyment, control appraisal and value appraisal were found to predict outcomes effectively, supporting the robustness of the SEM framework.

The study by Abualrob and Al-Saadi (2025) offers a theoretically grounded and empirically validated exploration of the factors influencing ICT integration in science education, highlighting the importance of both cognitive (TPCK, self-efficacy) and affective (enjoyment, control/value appraisal) dimensions. The comparative analysis between pre- and inservice teachers also offers useful implications regarding the importance of developing PSSTs' TPCK during their initial teacher education. With this noted, there are several limitations to the study: the study relies on self-reported questionnaire data that may be subject to bias; the cross-sectional design precludes conclusions about causality; and the study's participants were drawn from specific regions of Palestine, which may limit the generalizability of the findings.

It is worth noting that two of the four quantitative studies reviewed herein (i.e., Jeong et al., 2018; Sanchez-Martin et al., 2018) do not address or discuss matters of sampling, making room for new contributions that are well informed by specific quantitative designs and with clear stipulation of theoretical and methodological approaches. The most recent studies by Amponsah et al. (2024) and Abualrob and Al-Saadi (2025) address these limitations. Our knowledge about the role of emotions in preservice and inservice teachers' experience with technology would be greatly enhanced by more rigorous research approaches and a greater number of studies. Noting that four of the eight studies reviewed here did not identify theories of emotion, we also maintain that explicit and purposeful engagement with theory will enable theory testing when appropriate well-designed empirical research is carried out. More studies are needed that use both the open-ended and in-the-moment self-reporting methods in combination with observed interactions when research participants are engaging with a range of technologies and technological contexts, and a choice of methods that address a broader range of emotion labels beyond those reported through the included studies. Given the proliferation of digital learning tools and technologies, and the expansion of online learning, well-designed and theorised emotion studies in these science education contexts are needed.

7.3 Conclusion

This chapter reviewed a growing body of research examining the interplay between emotion and technology in preservice and inservice science teacher education. Across the eight studies reviewed, emotional outcomes related to online learning, instructional design, digital communities, and the integration of educational technologies revealing both opportunities and challenges for preservice and inservice science teacher development and professional practice. However, the body of work is marked by significant variability in theoretical orientation, methodological rigour, and interpretative depth.

Only three of the reviewed studies (i.e., Bellocchi et al., 2016; Amponsah et al., 2024; Abualrob & Al-Saadi, 2025) explicitly engage with emotion theories; namely, sociological perspectives (Collins, 2004; Turner, 2007), psychological perspectives (Pekrun, 2006), and emotional intelligence (Goleman, 1995). The remaining studies are atheoretical, often treating emotions as self-evident or merely descriptive, or authors utilizing their own conceptualizations such as *distance education emotions* or *emotional supp*ort.

Methodologically, the studies range from ethnographic case studies and phenomenological interviews to quantitative surveys and structural equation modeling. The more rigorous studies integrate validated instruments and clearly defined constructs, whereas others lack detail regarding design, sampling, and emotion measurement—particularly in the quantitative domain. Notably, four studies (Bellocchi et al., 2016; Cansoy, 2017; Eilon & Kliachko, 2004; Şevki Ayvaci et al., 2023) offer insights into emotional experiences through qualitative research designs and approaches that enable the collection of nuanced affective responses to online and technologically mediated instruction.

Empirical findings consistently indicate that emotional support—whether peer-based or instructor-facilitated—is central to fostering preservice and inservice science teachers' positive experiences in technology-rich environments. Emotional bonds and empathy among learners can alleviate isolation, particularly in online settings. However, unpleasant emotions such as anxiety, fear, and frustration often emerge in response to technological barriers, inadequate training, or lack of real-time interaction. The impact of pedagogical strategies on emotional experience is evident in studies on flipped learning (Jeong et al., 2018) and gamification (Sanchez-Martin et al., 2018), which report increased pleasant emotions, albeit without critical reflection on measurement validity or contextual influences.

Despite these contributions, a number of included studies do not address sampling rigor, validity concerns, or cultural specificity in emotion interpretation. The over-reliance on *post-hoc* questionnaires also risks oversimplifying emotional experiences in diverse technology-enhanced learning contexts. Although this body of research illuminates the ways in which emotions might shape preservice and inservice teachers' engagement with technology either through their professional learning or professional practice, more theoretically grounded and methodologically rigorous studies are needed.

References

NB Articles included in the systematic review are indicated using an asterisk (*)

*Abualrob, M. M. A., & Al-Saadi, S. H. (2025). Evaluating the role of technological pedagogical content knowledge in ICT-enhanced science education. *International Journal of Information and Education Technology, 15*, 117–126. https://doi.org/10.18178/ijiet.2025.15.1.2224

*Amponsah, K. D., Adu-Gyamfi, K., Awoniyi, F.C., & Commey-Mintah, P. (2024). Navigating academic performance: Unravelling the relationship between emotional intelligence, learning styles, and science and technology self-efficacy among preservice science teachers. *Heliyon, 10*(9). https://doi.org/10.1016/j.heliyon.2024.e29474

Barbe, W. B., Swassing, R. H., & Milone, M. N., Jr. (1979). *Teaching through modality strengths: Concepts and practices*. Zaner-Bloser.

*Bellocchi, A., Mills, K. A., & Ritchie, S. M. (2016). Emotional experiences of preservice science teachers in online learning: The formation, disruption and maintenance of social bonds. *Cultural Studies of Science Education, 11*, 629–652. https://doi.org/10.1007/s11422-015-9673-9

*Cansoy, R. (2017). Teachers' professional development: The case of WhatsApp. *Journal of Education and Learning, 6*(4), 285–293. https://doi.org/10.5539/jel.v6n4p285

Collins, R. (2004). *Interaction ritual chains*. Princeton University Press.

Durkheim, E. (1915/1965). *The elementary forms of the religious life* (Trans. J. W. Swain). Allen & Unwin.

*Eilon, B., & Kliachko, S. (2004). Perceptions of the teacher's roles by prospective elementary school science teachers in a web-based biology course. *Journal of Technology & Teacher Education, 12*(3), 339–360.

Goleman, D. (1995). *Emotional intelligence: Why it can matter more than IQ*. Bantam Books.

Goetz, T., Pekrun, R., Hall, N. C., & Haag, L. (2006). Academic emotions from a social-cognitive perspective: Antecedents and domain specificity of students' affect in the context of Latin instruction. *British Journal of Educational Psychology, 78*(4), 699–725.

*Jeong, J. S., Cañada-Cañada, F., & González-Gómez, D. (2018). The study of flipped-classroom for pre-service science teachers. *Education Sciences, 8*(4), 1630. https://doi.org/10.3390/educsci8040163

Mishra, P., & Koehler, M. J. (2006). Technological pedagogical content knowledge: A framework for teacher knowledge. *Teachers College Record, 108*(6), 1017–1054.

Özüdoğru, Ş. (2014). An essay on the role and importance of qualitative research in communication research. *Global Media Journal, 4*(8), 260–275.

Pekrun, R. (2006). The control-value theory of achievement emotions: Assumptions, corollaries, and implications for educational research and practice. *Educational Psychology Review, 18*(4), 315–341.

Petrides, K. V., Sangareau, Y., Furnham, A., & Frederickson, N. (2006). Trait emotional intelligence and children's peer relations at school. *Social Development, 15*, 537–547.

Rabi, R., Rahman, M. F., & Anwar, S. (2021). Emotional intelligence and academic achievement: A correlational study among undergraduate students. *Journal of Educational Sciences, 1*(1), 10–20.

Salovey, P., & Mayer, J. D. (2020). Emotional intelligence and academic success in college students. *Journal of Applied Psychology, 105*(5), 567–579.

*Sanchez-Martin, J., Cañada-Cañada, F., & Dávila-Acedo, A. (2018). Emotional responses to innovative science teaching methods: Acquiring emotional data in a general science teacher education Class. *Journal of Technology and Science Education, 8*(4), 346. https://doi.org/10.3926/jotse.408

*Şevki Ayvaci, H. Ş, Bebek, G., & Yamaçli, S. (2023). Experiences of pre-service science teachers in "teaching practice" during the sCovid-19 pandemic. *International e-Journal of Educational Studies, 13*, 134–152. https://doi.org/10.31458/iejes.1229871

Turner, J. H. (2007). *Human emotions: A sociological theory*. Routledge.

Venkatesh, V., Morris, M. G., Davis, G. B., & Davis, F. D. (2003). User acceptance of information technology: Toward a unified view. *MIS Quarterly, 27*(3), 425–478.

Wang, M. T., & Degol, J. L. (2017). Gender gap in science, technology, engineering, and mathematics (STEM): Current knowledge, implications for practice, policy, and future directions. *Educational Psychology Review, 29*(1), 119–140.

Chapter 8
Emotion and Science Teachers' Self-Efficacy and Identity

We identified 19 articles in the theme about self-efficacy ($n = 11$) and identity ($n = 8$) which dealt with science teachers' emotions (Akerson et al., 2014; Avraamidou, 2018; Can, 2015; Cooke et al., 2025; Cross & Hong, 2009; Hernández-Barco et al., 2021, 2022; Hobbs & Kelly, 2017; Kılınç et al., 2013; Koul et al., 2024; Mbowane et al., 2017; Melville et al., 2013; Mintzes et al., 2013; Polat et al., 2021; Rivera Maulucci, 2012; van Rooij, et al., 2019; Tang et al., 2025; Velasco et al., 2022; Yerdelen et al.,, 2019). The focus on emotions in these studies largely arose from theorizations of self-efficacy or identity, which acknowledged an affective dimension, however, three studies intentionally focused on and conceptualized emotion to understand science teacher identity (i.e., Avraamidou, 2018; Hernández-Barco et al., 2021, 2022).

In studies of self-efficacy and identity, emotions were predominantly an incidental focus due to the authors' theoretical choices. We begin with a discussion of emotion and theory in the self-efficacy studies, followed by a review of empirical outcomes. Then we present theories of identity and the role of emotions, followed by a review of empirical studies. In all cases, the presupposed role of emotion in teacher identity formation or change, and in self-efficacy was illustrated empirically. Disconfirming data were not presented, and theories were not questioned or revised considering empirical outcomes. In addition, idiosyncrasies of teachers' out-of-school circumstances were not considered as factors shaping the development of identity or self-efficacy, nor how personal lives afforded certain forms of instruction that were associated with emotion, and constitution of identity or self- efficacy.

A. Bellocchi et al., *Emotion Research in Science Teacher Education*, Sociocultural Explorations of Science Education 32, https://doi.org/10.1007/978-3-032-07711-0_8

8.1 Self-Efficacy Theory and Emotions

Theories of self-efficacy came predominantly from the discipline of psychology ($n =$ 7), with all studies drawing on Bandura's Social Cognitive Theory in which *affective state* forms one of four components of self-efficacy (Can, 2015; Hernández-Barco, et al., 2021, 2022; Mintzes, et al., 2013; Polat et al., 2021; Kılınç et al., 2013; Koul et al., 2024; van Rooij et al., 2019; Velasco et al., 2022; Yerdelen et al., 2019).

Two of the included studies provide conceptualizations of emotions replacing the element of affect derived from self-efficacy theory (Hernández-Barco, et al., 2021, 2022). Although these authors claimed to construct their own concepts of emotions, citing a range of recent Spanish or Turkish scholarship, the ideas presented largely reflect existing emotion theories such as Control Value Theory and classical ideas about stimulus–response dating to behaviorism. The studies do, however, afford greater power of analysis than what would be achieved if self-efficacy alone were used to understand emotions. In this way, the authors' conceptualizations enhance the focus on emotion and its associations with self- efficacy.

8.2 Empirical Outcomes of Emotion Research on Self-Efficacy

Studies of self-efficacy that reported emotional outcomes focused on three main themes: origins and development of self-efficacy; correlations between self-efficacy and other constructs (e.g., multiple intelligences, socio-scientific issues, mastery experiences, and STEM advocacy); and effects of self-efficacy interventions. Studies are discussed in this order. Overall empirical outcomes reflected the existing connection between emotional states and self-efficacy outlined in Bandura's theory. No studies suggested that the empirical evidence contradicted associations implied by theory.

Three studies report on the role of emotions in the origins or development of self-efficacy (Can, 2015; Koul et al., 2024; Polat et al., 2021; Yerdelen et al., 2019). Self-efficacy theory provides the basis for the role of emotional states in the development of self-efficacy, and the studies report on the sources of emotions. Can (2015) used semi-structured interviews to study the origins of self-efficacy beliefs of five female elementary preservice science teachers aged 21–25 from Turkey. Preservice teacher emotions were associated with self-efficacy beliefs derived from sources including observation of practicing teachers, and preservice teachers' personality traits and science content knowledge. When observing ineffective classroom management of undesirable classroom events, preservice teachers felt anxious, and this had a negative impact on self-efficacy. Emotionality related to self-perceived personality traits, such as shyness, nervousness, and emotionality, was associated with low self-efficacy. In contrast, the preservice teachers believed the ability to conceal emotion was efficacious. Low self-efficacy was associated with lack of science content knowledge and anxiety toward teaching incorrect information to students.

In a study of the role played by mentoring relationships in developing Thai elementary school teachers' self-efficacy as science teachers, Koul et al. (2024) address the role of emotions through culturally relevant constructs including *greng-jai* and *sanuk*. Greng-jai translates in English to mean "awe heart" or "fear heart," capturing a combination of reverence, respect, deference, homage, and fear that Thai people feel toward their seniors. Sanuk means intrinsic joy and pleasure and refers to fun. Using a sequential mixed-methods exploratory design (QUAN + qual) questionnaires administered to 834 teachers and interviews with six teachers, explored how teachers associated the terms *greng-jai* and *sanuk* with feelings about their mentors. Focusing on the role of emotion in teacher self-efficacy, negative associations were reported between greng-jai and participation of mentees and mentor in learning activities, whereas positive associations with sanuk were identified. Interviews revealed the negative impact of greng-jai on the mentor–mentee relationship and level of collaboration. In relation to the mentor- mentee relationship, emotional regulation, and emotional bonding (also connectedness) were reported. Emotions mediated the effect of peer collaboration and collegiality on self-efficacy as primary teachers of science. This is one of the few included studies that intentionally used culturally relevant emotions labels when studying a non-Western context.

Polat et al. (2021) studied self-efficacy of 189 Turkish preservice science teachers using a concurrent triangulation research design collecting data with a self-efficacy scale. Emotional state was the least important factor shaping self-efficacy beliefs. Unease, anger, and boredom were associated with having to respond to students' science questions. Anxiety was associated with teaching incorrect information and causing an accident that could harm them or their students.

Another study of Turkish final-year preservice teachers taught by the first author of the article aimed to investigate preservice science teachers' development of self-efficacy beliefs for teaching during a teaching practice course (Yerdelen et al., 2019). Participants were four female preservice teachers (Yerdelen et al., 2019). The teaching practice course involved micro-teaching and the authors investigated preservice teachers' experiences of the course using a case study design and self-efficacy questionnaires administered pre- and post-instruction. In relation to the emotional state element of self-efficacy, only one preservice teacher expressed pleasant emotions (relaxed, happy, motivated, enjoyment, self-confident) at the beginning of the course, whereas the other three expressed unpleasant emotions (worried, anxious, having difficulty). These unpleasant emotions were associated with communication skills, content knowledge, and classroom management skills. In contrast, following the course, all four teachers reported pleasant emotions in relation to instruction, classroom management, and communication. Two participants reported unpleasant emotions, one relating to content knowledge and the other to instructional skills.

There were four studies offering accounts of emotions associated with self-efficacy and constructs including multiple intelligences, socioscientific issues, mastery experiences, and STEM advocacy (Hernández-Barco, et al., 2021; Kılınç et al., 2013; Polat et al., 2021; Van Rooij et al., 2019). Hernández-Barco, et al. (2021) conducted a descriptive study to investigate correlations between self-efficacy, emotions, and multiple intelligences with a sample of 144 elementary preservice science teachers with different upper secondary backgrounds (i.e., humanities and social science or natural science, technology, and health) within the authors' university course in Spain. Questionnaires were constructed for each variable, which contained pre-determined emotion labels clustered into categories based on positive (pleasant) and negative (negative) valences, science content from the disciplinary combination of either Physics and Chemistry or Biology and Geology, and each of the multiple intelligences. Analyses produced associations between high self-efficacy, pleasant emotions, and logical mathematical and visual-spatial intelligences for both sets of disciplinary pairings. Naturalistic intelligence was linked with pleasant emotions toward Biology and Geology, but not with Physics and Chemistry, and there were no associations with self-efficacy. Neither gender nor multiple intelligences of preservice teachers predicted self-efficacy, with emotions and upper secondary background being the best predictors.

Genetically modified foods, a socioscientific issue, was at the center of Kılınç et al.'s (2013) study of Turkish preservice science teachers' efficacy beliefs. Using an exploratory mixed-methods design involving author-generated questionnaire about beliefs with 441 preservice science teachers from eight universities, and semi-structured interviews with eight preservice teachers, Kılınç et al. found connections between emotions and moral beliefs related to genetically modified foods. The results from quantitative and qualitative data reveal that preservice teachers associated embarrassment and guilt with preferencing of genetically modified foods over others. A few interviewees claimed that buying modified foods conflicted with their morals and emotions. These emotions were not associated with self-efficacy beliefs. Anxiety was the only emotion reported to be connected to a lack of self-efficacy in teaching about genetically modified foods because these topics were considered controversial by two interviewees.

Van Rooij et al. (2019) used questionnaires to address the lack of research on relationships between pleasant emotions, as a component of *emotional states* in self-efficacy, and mastery experiences with a sample of 69 Dutch university science students studying an education minor. Self-efficacy was predominantly related to mastery experiences and pleasant emotions, including enthusiasm, and feeling energetic, confident, and proud had higher self- efficacy. Unpleasant emotions were not related to self-efficacy.

In one of the three intervention studies, Hernández-Barco, et al. (2022) used a sample of 69 Caucasian elementary preservice teachers (52 men and 17 women, mean age = 22 years) from the authors' courses at a Spanish university. The preservice teachers' high school backgrounds were 80% from humanities, social sciences, and the arts, and 20% from science, health, and technology. The aim of the study was to determine preservice teachers' emotions associated with an intervention designed by the authors to teach about Integral Sustainability, and to determine the relationships between preservice teachers' emotions, content knowledge, and self-efficacy. An intervention with pre-test/post-test design involving two questionnaires designed by the authors was applied. The intervention unfolded over five phases which required preservice teachers to prepare instructional materials for elementary students and included a problem-based learning phase in which the preservice teachers designed and investigated a sand water-filter. Five aspects of self-efficacy correlated positively with emotions and science knowledge. Feeling more qualified to teach science was associated with high levels of satisfaction and surprise, and low levels of frustration. High levels of joy and confidence were associated with feeling more qualified to implement problem-based learning instruction. Efficacy about the use of future teaching methods was associated with emotions of confidence and satisfaction after completing the practical water-filter activity. Feeling surprised during the practical experiment phase correlated positively with increased motivation in science. Preservice teachers' sense of capabilities to conduct experiments had positive associations with high levels of satisfaction and low levels of boredom and rejection.

Continuing the focus on intervention studies, Mintzes et al. (2013) and Velasco et al.'s (2022) address the effect of interventions on self-efficacy that report associated emotional states. A mixed- methods non-equivalent control group design involving 116 elementary US teachers was used to study the effect of professional learning communities on self-efficacy related to science teaching (Mintzes et al., 2013) applied. In relation to the emotional states sources of self-efficacy, emotions associated with personal self-efficacy were reported. Fear was associated with childhood experiences of science, and anxiety over specific topics, such as fear of physics and love of life sciences, were also involved in self-efficacy beliefs. In relation to outcome expectancies, negotiating differences of opinion during collaboration in the design of science assessment involved forms of social persuasion associated with feelings of emotional support.

Velasco et al. (2022) used an embedded single case-study design with multiple methods to study the development of self-efficacy of STEM teacher leaders in advocacy for STEM education. The first author was included as a participant in the study, which consisted of 10 leads who formed part of a STEM ambassador program. Analysis of textual data revealed that the *emotional states* component of self-efficacy was the least frequently occurring code. Instances involving emotion involved anxiety about beginning to advocate for STEM in K-12 education. Of the four elements constituting self-efficacy theory, the emotional state code appeared 12 times out of 157 codes and represented the least frequently (8%) occurring component of self-efficacy. There were only 22 codes out of 157 related to pleasant emotions connected with improved self-efficacy, and anxiety (unpleasant emotion) was mentioned in relation to beginning to advocate for STEM education but subsided with experience.

8.3 Identity Theory and Emotions

Studies focusing on science teacher identity present a diverse range of theoretical and conceptual approaches. In five studies, emotions arose from the identity theory that was applied (professional identity, Akerson et al., 2014; Mbowane et al., 2017; ecological perspective, Rivera Maulucci, 2012; post-structuralist perspectives, Melville et al., 2013). One study drew on emotion theory or conceptualizations of emotion in conjunction with identity theory (achievement emotions, Avraamidou, 2018), whereas another drew on affect-as-information, which includes emotions as sources of information for the self (i.e., Tang et al., 2025). Two studies were unique in that one proposed a conceptual model, Beliefs Identity Goals Emotion, which addressed emotions (Cross & Hong, 2009), and in the other, emotions were associated with conceptualization of aesthetics (Hobbs & Kelly, 2017). In one study, emotional work is the construct used to explain teacher identity in a professional learning community of academics and classroom teachers who are seeking to enact justice-centered science instruction (Cooke et al., 2025). No theoretical or conceptual orientation to emotion work is foregrounded, however, the authors drew on concepts from other research on science teacher identity when discussing emotions, suggesting an atheoretical approach to emotions (cf. Chubbuck & Zembylas, 2008; Rivera Maulucci, 2013).

The theoretical/conceptual diversity presented by these studies makes synthesis of studies and their findings challenging as conceptualizations of emotion can differ markedly, reflecting a trend in all themes presented so far in this book. Any one of these theoretical/conceptual traditions could be continued in future research to extend knowledge developed from the studies included here.

8.4 Empirical Outcomes of Emotion Research on Identity

The topics of research on science teacher identity which involve emotions range from self- study of a professor becoming an elementary teacher of Nature of Science (Akerson et al., 2014), identity development during science education reform from direct instruction to science inquiry (Cross & Hong, 2009; Melville et al., 2013), developing teacher knowledge, identity, and passion for science (Hobbs & Kelly, 2017), the role of science fairs in teacher identity development (Mbowane et al., 2017), connections between identity and emotions when teaching for social justice in science education (Cooke et al., 2025; Rivera Maulucci, 2012), and modeling of interplay between emotion, identity, and personal epistemologies and epistemic stances (Tang et al., 2025). As the topics suggest, there are studies addressing specific kinds of science teacher identity, such as becoming a Nature of Science teacher, social justice teacher, or a science inquiry teacher (e.g., Akerson et al., 2014). Whereas others are about general science teacher identity development (e.g., Hobbs & Kelly, 2017).

Overall, empirical outcomes across the studies draw attention to emotion in relation to changing teacher belief systems, confirming beliefs, nature of curriculum and reform about science inquiry, interpersonal matters (students or colleagues), pedagogies, changing identity as science teachers, clarifying the role of the science teacher, sense of efficacy as a teacher of science, and the overall teaching experience. Changes associated with these elements are sources of anticipatory emotions, experiential emotions, and these elements can also be shaped by science teachers' emotions. Theories of identity present the relationship between emotions and identity formation or affirmation, and empirical findings reflect these theoretical connections, with no study providing empirical outcomes contradictory to the underlying assumptions.

Two studies focus on science teacher identity in relation to science inquiry, addressing emotions associated with forming identities as *teachers of inquiry* during curriculum reform (Cross & Hong, 2009; Melville et al., 2013), whereas one study focuses on the first author's journey in becoming a *teacher of Nature of Science* (Akerson et al., 2014). The study by Melville et al. (2013) constructs *direct instruction* as a dominant discourse in inservice and preservice science teaching respectively, which is countered as the teachers form identities as *teachers of inquiry*. Emotions are evoked and involved in identity formation; a perspective also present in Akerson et al. (2014).

Melville et al. (2013) focus on the authors' teaching of preservice science teachers, in which scaffolded science inquiry activities are central parts of instruction. This instruction responds to Canadian curriculum reforms which emphasized inquiry. Replicating the narrative methodology and interview methods of their previous study (see Chap. 5), the article presents outcomes relating to 31 preservice teachers across classes from three consecutive calendar years. Findings focus on the way in which the scaffolded instruction serves to constitute preservice science teachers' identities as teachers of inquiry. Once again, the analysis focuses on the way in which discourse experience and emotion are important for identity formation. Emotions arise in relation to the preservice teachers' pleasant experiences with the scaffolded inquiry activities, leading to positive dispositions towards inquiry. Open forms of inquiry were associated with struggle, through which the preservice teachers persisted. Through the poststructuralist lens adopted, the authors concluded that these emotions were important for identity formation.

The final study on science teacher identity and inquiry in curriculum reform is concerned with the impact of reform on inservice science and mathematics teachers' emotional experiences (Cross & Hong, 2009). Without an explicit research design being presented, five science teachers and one mathematics teacher from the USA form the sample. Using a Beliefs Identity Goals and Emotions conceptual framework created by the authors, the article reports experiential trajectories of the teachers through descriptions of four paths labelled A-D. In a similar way to the studies using a poststructuralist perspective on identity, Cross and Hong's (2009) framework assumes that identity forms from experiences, which gives rise to emotions and, in turn, shape identity. Paths A and C of identity development illustrate how pleasant emotions shape identity. In Path A, pleasant emotions arise when teachers adopt and interpret reform mandates based on their beliefs and identities, whereas in Path

C pleasant emotions arise during implementation of classroom goals in specific classroom contexts. Paths B and D were associated with unpleasant emotions. Path B reflected the way in which unpleasant emotions arose when setting instructional goals requiring negotiation between existing teacher beliefs and identity with the practices necessary for implementing the reform agenda. In contrast, Path D captures the way teachers' implementation of reform-oriented goals lead to unpleasant emotions.

Focusing on the first author's experience in a 3rd-grade class, Akerson et al. use video data, a teacher-log, and student work to understand Akerson's professional identity development as a teacher of Nature of Science. The authors present their own conceptual framework on professional identity. An element of this framework identifies emotions within a theme named Personal struggle and activity. Outcomes reveal that pleasant, anticipatory emotions (nerves, excitement, and happiness) are associated with teaching Nature of Science. Whereas the pleasant feeling of connectedness also arises from interactions with students and colleagues. Unpleasant emotions (frustration, feeling drained) revolve around the inability to meet student needs, especially those with diverse needs. These unpleasant emotions stem from initial, or antecedent, emotions of care for students. Other unpleasant emotions arise when the teacher questions her competency. Despite the mix of pleasant and unpleasant emotions, early experiences recorded in the teacher log suggest that unpleasant emotions act as threats to the goals of teaching Nature of Science and forming an identity as a teacher of Nature of Science. Ultimately, interpretation and reinterpretation of self through external factors and others' perceptions, as well as the influence of sub-identities, are all important in Akerson seeing herself as a teacher of Nature of Science.

In contrast to reforms about inquiry and Nature of Science teaching and learning, Mbowane et al.'s (2017) study addresses teacher emotions and identity in the context of science fairs, which can require students to engage in project and practical work. Science fairs, in South Africa, are seen as important ways of improving innovative and gifted students' academic performance and skills in science and mathematics. Adopting a conceptual framework on teachers' professional identity and roles in science fairs, emotions arise as one element which is part of the process of identity formation. Using a mixed-method (i.e., QUAL-quant) design within interpretive and descriptive paradigms, the study focuses on interviews with a sample of 10 Physics teachers from two schools. Seeking to understand the educational significance of fairs and the extent to which they provide professional development for the teachers. One theme in their findings addresses emotions. The authors report that teachers experienced emotions associated with student interactions, student achievement, and when assisting students with fair projects. These emotions were part of a process which involved overall growth in teachers' sense of professional identity.

One study investigated the links between emotions and science teacher identity based on general aspects of science teaching (Rivera Maulucci, 2012). Drawing on data from two preservice and one inservice science teacher from the USA, Rivera Maulucci (2012) applies an ecological framework on emotions and identity and a narrative inquiry design. The study draws on a diverse array of data from sources including classroom observations, semi-structured life history and phenomenological interviews, lesson plans, seminar coursework, and video data from college and K-12 classroom settings. Key outcomes include that teachers' emotions frame their sense of self, influence teaching goals, are indicative of teaching experiences, relate to beliefs about science, create a sense of self-efficacy about science or science teaching. In addition to the role of their own emotions, teachers also drew on *student emotion* to affirm their teaching identity, and their views of science as a collaborative exercise, which creates positive affiliation with an identity as a scientist. External verification of one's identity as a science teacher from peers or students was also associated with emotions, through which positive affirmations helped sustain their science teacher identity.

Over one year, Cooke et al. (2025) conducted a qualitative study using video recordings to capture 21 professional learning community meetings established between the researchers and 13 inservice science teachers in the US. The goal was to investigate how professional learning communities can support the development of teacher identity towards practicing social justice in science classrooms. Three key findings were reported, with one establishing connections between emotional work of discussing race and spurring one another on while maintaining accountability to achieve social justice practices. Participating teachers reported the fear of facing classroom discussions about race and injustice. Support provided by members of the learning community meant that fear did not become a deterrent for enacting social justice pedagogy.

Whereas the two preceding studies treated emotion and identity as discrete constructs, Tang et al. (2025) remain agnostic to towards frameworks that consider these and epistemic constructs (epistemic stance, personal epistemology) as discrete versus approaches that treat these constructs as a unified aspect of human experience. Their interest is to use data from an argument that broke out between a supervising classroom teacher and a PSST intern during instruction when completing teaching placement, to model the a priori assumption that affect, identity, and epistemic stance are co-constitutive. An epistemic stance is a person's contextually bound, expressed or enacted, orientation towards knowledge, knowing, and learning. The classroom argument centered on the differences of opinion between the two teachers about the level of inference year eight students should use to interpret results from scientific investigations they had completed. The inservice teacher held the view that no inferences should be made, whereas the PSST argued that empirical evidence should be combined with students' prior knowledge about phenomena when making claims.

A range of data including video recordings of the argumentation exchanges, interviews, and lesson artifacts (e.g., lesson plans, resources) were used to model what the co-constitution, or entanglements, of affect (including emotion), identity, and epistemic stances. The authors state that no specific analytic codes were used to identify affect, identity, and epistemic stances due to the focus on proposing high-inference explanations of entanglements instead of reporting empirical findings. They report the use of identity markers and instances of affect and epistemic stances. In analyzing the data, markers and instances correspond to specific enactments. For example, an utterance such as *I am a people person,* is taken as a marker of identity in the data. A teacher stating they are *worried about closing off student ideas* is an instance of affect. When a teacher states that cause-effect relationships are fundamental to science, this is an epistemic stance marker. Overall, the supervising inservice science teacher is represented as having a *things-person identity* and a *purist identity,* which means a focus on science over a humanist focus on students. In contrast, the PSST is constructed as having a *people-person* identity, which reverses the formula reported for the inservice teacher.

Noting how measured the debate in class was, Tang et al. note it was very difficult to find instances of affect, and therefore, emotion, from video data. However, using self-reported interview data, the authors report high-level inferences about the inservice teacher's *love for the eternal* (e.g., laws of science), and her identity as a purist. This emotion-infused identity is associated with her focus on having students in the class make empirical claims, and avoiding inferences (i.e., her epistemic stance). For the PSST, *worry of shutting down student ideas* was the emotional component of a *people-person* identity, which, in the context of the classroom argument, manifest as an epistemic stance that values student contributions and engagement. In this way, the PSST's epistemic stance reflects an epistemology of learning, as opposed to the inservice teacher's epistemology of science. Tang et al. conclude that a researcher's choice of modelling identity, affect, and epistemic stance as discrete constructs or as co-constitutive elements that activate concurrently has different implications for the ways in which professional learning and science teacher education is structured. Although the study considered emotion to be a part of affect, a specific focus on emotion was difficult to discern from more general ideas about affect.

The last two studies included in this section focus on identity and emotional trajectories (i.e., Avraamidou, 2018; Hobbs & Kelly, 2017). In Southern Europe, Avraamidou (2018) reports the lived experiences of an elementary inservice teacher using a life-history biographical research design. Lived experiences were documented over one year using methods that included a science-biography, three brief biographical assignments, a personal philosophy of teaching statement, and an interview and interpreted through a framework that included identity, achievement emotions, and *perezhivanie*, which describes how people perceive, experience, and process the emotional aspects of their lives. These diverse methods captured the teachers' perceptions and emotions of her own school experiences in which science in elementary school had strong connections to the teacher's personal background, but then became boring in secondary school, which impacted her performance on assessment. In contrast, during her science methods course, taught by the author of the article, the teacher formed a *science identity* because emotions of confidence emerged due to the support provided with learning to teach science. Feelings of comfort and enjoyment arose and are constructed to have influenced her competence. In this way, her identity as an elementary science teacher shifted through an arc of pleasant (elementary), unpleasant (secondary), to pleasant (teacher education course) emotions across the teacher's lived experience of science. The findings of the study supported the connection between emotion and identity formation which was present in the identity framework informing the study.

Hobbs and Kelly (2017) report on the experiences that evoked strong emotional and cognitive outcomes shaping teacher knowledge, professional identity, and passion for science education. The sample from Australia includes a specialist elementary teacher, two preservice elementary teachers, and one marine educator. The authors present their own knowledge-identity-passion framework to describe the teachers' experiences. Central to the framework are cognitions, emotions, and actions. Findings are generated through application of the framework on teacher narratives that were obtained via interviews in three separate studies. As with previous studies on identity and emotion, Hobbs and Kelly find that affirmation from colleagues and students' learning and excitement about science serve to form and affirm the teacher's identities as teachers of science.

8.5 Conclusion

Nineteen studies were included in the theme science teacher emotions and self-efficacy ($n = 11$) and identity ($n = 8$). Overall, most studies addressed preservice teacher emotions more frequently than inservice teacher emotions. Emotion was a focus in this body of work largely due to emotion or affect (used synonymously with emotion) being components of theories of self-efficacy or identity. In four cases, emotions represented a stand-alone focus, and the study adopted emotion theory to define and conceptualize emotion in relation to teacher identity formation (i.e., Avraamidou, 2018; Hernández-Barco et al., 2021, 2022; Tang et al., 2025). Two studies developed a conceptual framework that included an emotion emphasis (cf. Cross & Hong, 2009; Hobbs & Kelly, 2017). Intentional focus on emotion defined through emotion theory provided effective granularity, which meant that these studies offered specific insights into emotion and identity formation or self-efficacy development than those relying on identity and self-efficacy theories alone, which capture emotion through the omnibus *affect* construct. Only one study considered emotion constructs that were culturally relevant to the sample studied (i.e., Koul et al., 2024). There was greater theoretical homogeneity in the self-efficacy studies, which drew on Bandura's theory of self-efficacy, when compared to the theories of identity used in the identity studies. Overall, studies empirically supported the connections between emotion and identity, or self-efficacy established by the identity and self-efficacy theories or frameworks adopted by the authors.

Studies of self-efficacy reported on the role of emotions in origins and development of self-efficacy, correlations between self-efficacy and other constructs (e.g., multiple intelligences, socio-scientific issues, mastery experiences, and STEM advocacy) and effects of self-efficacy interventions. Key empirical findings about science teacher emotion and self-efficacy point to factors relating to science teachers as individuals (e.g., personality traits; science content knowledge; intelligences; prior science background; gender), relational factors (e.g., mentor–mentee relationships; teacher-student), curriculum (e.g., topics- genetically modified food; science subject-physics), pedagogy (e.g., experiments, practical activities), assessment, and development (e.g., teaching observations; professional development). Emotions, predominantly nested within *affect*, broadly serve as antecedents, outcomes, or part of the process of developing self-efficacy. Three studies reported negative or low associations between emotions and self-efficacy (cf. Hernández-Barco, et al., 2021; Van Rooij et al., 2019; Velasco et al., 2022).

Regarding science teacher emotion and identity, researchers report the role of emotions and specific kinds of science teacher identity (Nature of Science teacher, social justice teacher, science inquiry), general science teacher identity development, and the impact of specific experiences on emotion and identity. As with self-efficacy, a range of factors shape the formation of teacher identity and the associated role of emotion. Emotions can be both antecedents of identity formation, outcomes of it, and aspects of formative experiences that lead to identity formation. Emotions and identity are associated with pedagogy (Nature of Science, science inquiry, social justice), personal factors (intelligences; beliefs, personal goals), interactional factors (e.g., mentoring; teacher-student interaction), pathways of development, teachers' past school experiences as students, curriculum reforms, and external validation (e.g., student feedback; peer affirmation). Synthesis of empirical findings was challenging due to the heterogeneity in identity theories, which provides for different constructs informing the studies reviewed.

As a general conclusion, despite conceptual divergences, studies show the importance of emotion as antecedents, outcomes, or part of the self-efficacy and identity formation experiences of science teachers.

References

NB Articles included in the systematic review are indicated using an asterisk (*)

*Akerson, V. L., Pongsanon, K., Weiland, I. S., & Nargund-Joshi, V. (2014). Developing a professional identity as an elementary teacher of nature of science: A self-study of becoming an elementary teacher. *International Journal of Science Education, 36*, 2055–2082. https://doi.org/10.1080/09500693.2014.890763

*Avraamidou, L. (2018). Elementary science teacher identity as a lived experience: small stories in narrative analysis. In P. A. Schutz et al. (eds.), *Research on teacher identity* (pp. 145–155). Springer. https://doi.org/10.1007/978-3-319-93836-3_13

*Can, H. (2015). Sources of teaching efficacy beliefs in pre-service science teachers. *Elementary Education Online, 14*(3), 1044–1059.

Chubbuck, S. M., & Zembylas, M. (2008). The emotional ambivalence of socially just teaching: A case study of a novice urban schoolteacher. *American Educational Research Journal, 45*(2), 274–318. https://doi.org/10.3102/0002831207311586

*Cooke, H., Campbell, T., Luehmann, A., Zhang, Y., & Scipio, D. (2025). "I sit here thinking I can do this"—Developing justice-centered ambitious science teaching identities in professional learning communities. *Journal of Research in Science Teaching, 64*(2), 1040–1072. https://doi.org/10.1002/tea.21991

*Cross, D. I., & Hong, J.-Y. (2009). Beliefs and professional identity: Critical constructs in examining the impact of reform on the emotional experiences of teachers. In A. Schutz & M. Zembylas (Eds.), *Advances in teacher emotion research.* Springer. https://doi.org/10.1007/978-1-4419-0564-2_14

*Hernandez-Barco, M.A., Sanchez-Martín J., Corbacho-Cuello I., & Cañada-Cañada F. (2022). Is sustainability an exhausted concept: Bridging the gap from environmental awareness to emotional proficiency in science education through integral sustainability. *Heliyon, 8*(12). https://doi.org/10.1016/j.heliyon.2022.e12546

*Hernández-Barco, M., Cañada-Cañada, F., Corbacho-Cuello, I., & Sánchez-Martín, J. (2021). An exploratory study interrelating emotion, self-efficacy and multiple intelligence of prospective science teachers. *Frontiers in Education, 6*, Article 604791. https://doi.org/10.3389/feduc.2021.604791

*Hobbs, L., & Kelly, L. (2017). The heart of the educator: Aesthetic experience shaping knowledge, identity, and passion. In A. Bellocchi, C. Quigley, & K. Otrel-Cass (Eds.), *Exploring emotions, aesthetics and wellbeing in science education research.* Springer. https://doi.org/10.1007/978-3-319-43353-0_4

*Kılınç, A., Kartal, T., Eroğlu, B., Demiral, Ü., Afacan, Ö., Polat, D., Demirci, G., Mutlu, P., & Görgülü, Ö. (2013). Preservice science teachers' efficacy regarding a socioscientific issue: A belief system approach. *Research in Science Education, 43*, 2455–2475. https://doi.org/10.1007/s11165-013-9368-8

*Koul, R., Musikul, K., Nishikawa, S., & Almutawa, H. (2024). Mentoring in the development of science teaching self-efficacy among primary school teachers in Thailand: A Mixed Methods study. *Journal of Science Teacher Education, 35*(1), 44–62. https://doi.org/10.1080/1046560X.2023.2179384

*Mbowane, C. K., de Villiers, J. J., Rian, B., & Max, W. H. (2017). Teacher participation in science fairs as professional development in South Africa. *South African Journal of Science, 113*(7/8), 7. https://doi.org/10.17159/sajs.2017/20160364

*Melville, W., Bartley, A., & Fazio, X. (2013). Scaffolding the inquiry continuum and the constitution of identity. *International Journal of Science and Mathematics Education, 11*, 1255–1273. https://doi.org/10.1007/s10763-012-9375-7

*Mintzes, J. J., Marcum, B., Messerschmidt-Yates, C., & Mark, A. (2013). Enhancing self-efficacy in elementary science teaching with professional learning communities. *Journal of Science Teacher Education, 24*, 1201–1218. https://doi.org/10.1007/s10972-012-9320-1

*Polat, D., Uluay, G., & Başarmak, U. (2021). Examining the pre-service teachers' components of self-efficacy beliefs in science teaching (SEBST). *Sakarya University Journal of Education, 11*(1), 195–217. https://doi.org/10.19126/suje.798077

*Rivera Maulucci, M. S. (2012). Exploring linkages between identity and emotions in teaching for social justice in science teacher education. In M. Varelas (Ed.), *Identity construction and science education research: Learning, teaching, and being in multiple contexts* (pp. 123–139). Sense Publishers. https://doi.org/10.1007/978-94-6209-043-9_9

*Rivera Maulucci, M. S. R. (2013). Emotions and positional identity in becoming a social justice science teacher: Nicole's story. *Journal of Research in Science Teaching, 50*, 453–478. https://doi.org/10.1002/tea.21081

*Tang, X., Elby, A., Levin, D. M., & Chumbley, A. K. (2025). Complex dynamics behind the scene: How identity and affect shape epistemic stance in science teaching. *Cognition and Instruction, 43*, 134–173. https://doi.org/10.1080/07370008.2025.2473070

*van Rooij, E., Cornelia, M., Fokkens-Bruinsma, M., & Goedhart, M. (2019). Preparing Science undergraduates for a teaching career: Sources of their teacher self-efficacy. *Teacher Educator, 54*(3), 270–294. https://doi.org/10.1080/08878730.2019.1606374

*Velasco, R. C. L., Hite, R., & Milbourne, J. (2022). Exploring advocacy self-efficacy among k-12 stem teacher leaders. *International Journal of Science and Mathematics Education, 20*, 435–457. https://doi.org/10.1007/s10763-021-10176-z

*Yerdelen, S., Osmanoglu, A., & Tas, Y. (2019). The influence of a teaching practice course with video-case enriched microteaching on prospective teachers' self-efficacy for teaching. *International Journal of Research in Education and Science, 5*(2), 560–573.

Chapter 9
Emotion and Science Teacher's Work

It seems like a platitude to say that the nature of science teachers' work is multifaceted and thereby presents numerous avenues that connect with emotions. The emotional exhaustion commonly associated with teaching—a key contributor to teacher wastage from the profession, which is often gendered—is well documented (cf. Aloe et al., 2014; Olson et al., 2019; Weldon, 2018). Despite this issue being common knowledge, the problem persists and for the science, technology, engineering, and mathematics disciplines, it poses a threat to the low supply and high demand for suitably qualified schoolteachers globally. Researchers have raised questions about the level and quality of preparation of science teachers for the emotional work that is an inevitable part of the job and questions about the structural factors that intensify this work (Bellocchi, et al., 2018b; Zembylas, 2004). Unsurprisingly then, we find that this theme is the most voluminous in the number of published articles.

Included papers in this theme ($n = 29$) are almost equally distributed across school settings with 15 studies involving elementary teacher samples, 10 studies clearly stating middle school or high school samples, and a further 4 involving high school samples by inference. Reviewed studies have been categorized in terms of the work of inservice teachers as distinct from the work of preservice teachers based on the inherent differences in these two contexts, for invoking different emotional demands. Empirical findings of research examining the work of inservice science teachers highlight the role of emotion in relation to pedagogical content knowledge development, decisions related to managing a science classroom, including the role of emotions in relation to ongoing professional learning including the identification of antecedents of teacher emotion, the emotions associated with changing careers, working in a trauma-informed manner, and fostering creativity and enacting bilingual science education. The empirical findings of research examining the work of preservice teachers includes distinct, yet related foci: emotion associated with learning to teach science, broadly, and within distinct science disciplines, more specifically; and

A. Bellocchi et al., *Emotion Research in Science Teacher Education*, Sociocultural Explorations of Science Education 32, https://doi.org/10.1007/978-3-032-07711-0_9

the role of emotion in teacher identity formation, particularly in relation to marginalization and issues of social justice. Before presenting the empirical findings, the theories of emotion that underpin the studies on inservice teachers' work, as distinct from preservice teachers' work are described.

9.1 Theories of Emotion in Research Examining Inservice Teachers' Work

Eighteen of the reviewed papers examine emotions and the work of inservice teachers, with the majority of studies adopting psychological perspectives: Kordíková and Brestenská (2022) draw on the theory of semantic differentials to gauge respondents' subjective attitudes on the emotional satisfaction associated with bilingual science teaching; dos Santos and Mortimer (2003) categorize primary and secondary emotions through observations of physiognomy and intonation to examine how the affective and emotional elements of teacher-student interactions contribute to positive or negative attitudes towards chemistry; Bullough (2009) draws on Appraisal theory/identity (Solomon, 1993, 2004) to examine how emotions organize perception, cognition and behavior and contribute to the development of a person's identity. Three further papers draw on another psychological theory—Control Value Theory (CVT) (Pekrun, 2006)—Membiela et al. (2023) recognize emotions as an affective arousal mechanism associated with tasks like studying or results such as success and failure (cf. Pekrun & Perry 2014); Uzuntiryaki-Kondakci et al. (2022) supplement CVT with the process model of emotion regulation (Gross, 1998) to examine science teachers' emotions, and emotion regulation goals and strategies during instruction, alongside the role of teaching experience, teacher efficacy beliefs, and teacher goal orientations in their emotions and emotion regulation; and Thomson and Turner (2019) who examine the proposition that within education settings, one's emotions and motivations are interrelated. One study, Brick et al. (2021), adopts a neurobiological perspective to examine how learning about neuroscience affects teachers' own emotional or affective states, which become central to the process of encouraging and optimizing student learning. A collection of four papers (Zembylas, 2001, 2002, 2004a, 2004b) take a multidisciplinary lens to the theory of emotion, and another study constructs a conceptual framework (Mahmood et al., 2024). Two of the studies adopt a sociological perspective (Bellocchi et al., 2018a; Ritchie et al., 2011). Three papers are atheoretical (i.e., González-Weil et al., 2025; Hong & Kang, 2010; Praderio Gaias et al., 2024).

9.2 Theories of Emotion in Research Examining Preservice Teachers' Work

Although most studies related to inservice teachers' work deploy a psychological theory of emotion, the studies related to the work of preservice teachers more multi-disciplinary approaches. Of the three papers drawing on a psychological approach, Wicaksono and Korom (2023) approach emotion through the framework of three dimensions of attitude towards science where emotion is considered part of the affect element of attitudes. Erdemir and Kandil İngeç (2023) define emotional empathy as the sharing of emotional responses whereas cognitive empathy is the ability to accurately recognizing and ascribing emotions to others. Ma (2018) draws on Krathwohl et al.'s (1964) three domains of learning (cognitive, affective, & psychomotor) to acknowledge that affective phenomena occur on different levels, then elaborates by drawing on research on interest to categorize those phenomena into *habitual attributes*, which result in people reengaging with content over time or readily activate certain emotional states, and *affective experiences*, which are elicited by environmental stimuli.

Four of the papers take a multi-disciplinary view of emotion: Rivera Maulucci (2011) view emotions as indicators in response to a change in the status of goals and consider how more intense emotions indicate more specific changes in an individual's response. Brígido et al. (2013) define emotions as reactions to information from the environment, including relationships while recognizing that the intensity of an emotion is mediated through subjectivity including culture and gender. This study categorizes emotions as positive (i.e., pleasant), negative (i.e., unpleasant), or neutral. Brígido et al. (2013) include additional theory to state that emotions can be induced by the recall of past events and may also involve the anticipation of future events, and that emotions are central to decision-making. This theoretical work is elaborated in Borrachero et al. (2014) by highlighting that rewards and punishments (i.e., instrumental reinforcers) can induce emotions.

Three of the papers concerned with preservice teachers' work and emotions adopt an approach to emotion informed predominantly by Education scholarship. Zembylas and Buhlman Barker (2002) contribute to and extend upon research related to studies of inservice science teacher education by examining preservice teachers' reflections, emotions, and attitudes as informed three dimensions of learning by highlighting the importance of cognitive and affective domains to teaching and learning and then draw on Zembylas (2001, 2002, 2004a) to argue for the role that emotions play in teachers' work including the development of pedagogical content knowledge, planning practices, and relationships with students, parents, and colleagues. Zembylas and Buhlman Barker (2002) refer to the *emotional ecology* of a teacher which is connected to different aspects of their pedagogical content knowledge across levels including individual, relational, and socio-political spheres. Rivera Maulucci (2013) draws on the concept of *emotional navigation* and *emotional ambivalence* to refer to the ways in which individuals actively manage and express their emotions in relation to actions. They apply the concept of *critical emotional praxis* to acknowledge that emotions play a key role in challenging and transforming injustice. Reference is also made to Solomon (2007) when acknowledging that emotions not only play a role in teachers' school lives, but also in the ways that emotions contribute to teacher identity.

The paper by Ritchie et al. (2007) draws on identity theory (Flores & Day, 2006) to acknowledge that an individual's identity is mediated by discursive practices (i.e., cognitive and emotional practices) between self and others in the field that have cultural antecedents. Moore Mensah (2019) offers no theorization of emotion but draws on critical race theory to state that race plays a significant role in the emotional experiences of preservice teacher candidates as they work to shape their identity as a teacher.

Overall, a more diverse range of theory has been applied in studies of preservice science teachers when compared with inservice science teacher research.

9.3 Empirical Outcomes on Emotion in Relation to the Work of Inservice Science Teachers

9.3.1 Emotions and Science Teaching Experiences

We begin this section with a review of interrelated articles by Michalinos Zembylas, before discussing other included work. Zembylas (2001) presents an essay inspired by a three-year ethnographic study conducted with Catherine Myers, a 25-year veteran elementary school teacher, which reveals how emotions play significant roles (both implicitly and explicitly) in Catherine's development as a science teacher. Zembylas seeks to reclaim the idea that emotions and cognitive phenomena are interrelated. Framed in this understanding of emotion and cognition, Zembylas (2001) presents analysis of two *vignettes* of experience which are theorized as emotional genealogies of teaching that seek to record events through time in instances that seem "without history—in sentiments, love, conscience, instincts" (p. 115).

The first vignette describes how Catherine saw her students being excited, which affected not only her emotions (delighted, fascinated, excited), but also her future decisions and actions: "As Catherine said elsewhere during our discussions, her excitement seemed to nurture the children's fascination and vice versa, depending on the activity and the circumstances. But there was a mutual support of one another that created and sustained their pleasant emotions and attitudes" (Zembylas, 2001, p. 118). Then, in the second vignette, the focus shifts to explore the role that Catherine's emotions played in the development of *her* self-confidence in science teaching. In the early years of her career, Catherine felt guilt and disappointment because she did not feel as though she was being an effective science teacher. Initially, she blamed herself for these perceived inadequacies but later realized that many factors were contributing to her feelings including the philosophy of the school system which emphasized content knowledge and standardized testing as well as what she perceived to be the masculinized character of science—objective and rational being privileged through that system. She also experienced frustration due to concerns expressed by colleagues that her students would lack content knowledge due to her student-centered teaching approach. She felt dismissed, which undermined her self-confidence and self-esteem, leading her to question her teaching approach and her values. These feelings shifted when Catherine worked in the kindergarten context where she observed a freedom with feelings; in particular, her students' fascination and curiosity with their surroundings, which reminded her of her own feelings as a child. As a result of those experiences, Catherine reached the view that "emotions are ways with which we know the world around us" (p. 119). Zembylas (2001) concludes that the findings of his work invite reflections on the interrelationships between teachers' work and their emotions and students' experiences of science education.

Data from Zembylas' (2001) ethnographic study formed the basis for a subsequent journal article (Zembylas, 2002). In the journal article, Zembylas (2002) introduces notions such as *emotional geographies of schooling* to describe experiential patterns in human interactions, and *heightened emotionality* to describe the emotional elements of teaching practices across contexts.

The article deploys the emotional genealogy of teaching as theorized in Zembylas (2001) as the methodological approach to data analysis and adopts an analytic framework that recognizes the role of teachers' emotion three levels: (1) intrapersonal level—this level is concerned with how science teachers express emotions on an individual/personal level; (2) interpersonal level—refers to how science teachers use their emotions in their relationships with others; (3) intergroup level—refers to the interrelationships between science teachers' emotions and the socio-cultural context within which they work. Insights derived from these three levels include how the genealogy of emotions made visible the process that led to Catherine's development of habitual tendencies; the impact Catherine's emotions had on her pedagogical decision-making and the context for these decisions; the expression of Catherine's emotions suggest that some prevalent emotional attitudes underpinned her pedagogical practices; and, finally, Catherine's emotions were driving the social relations, values, and practices that were both realized and symbolized in the context of science education in her setting. The insights shape a call for recognition of the emotional labor involved in science teaching as teachers navigate their own emotions, as well as those of their students and colleagues inside a broader institutional context. In addition, Zembylas (2002) argues that Catherine's case unveils the affective side of teaching and the power of emotions to transform teachers' experiences, suggesting that the analysis demonstrates that emotions can enable science teachers to think and act differently. The implication for this work is that both the pleasant and unpleasant emotions of science teaching need to be explored to improve science teachers' work environments and to understand the emotional rules that limit pedagogical innovation.

The same ethnographic case-study from 2001 was used to explore the use of *emotion metaphors* used by Catherine to conceptualize and enact her work (Zembylas, 2004a). Qualitive data were analyzed using a wholistic approach and a selective approach in that the narrative was reviewed and, where emotion metaphors were apparent, the metaphors and the context of the metaphor usage were analyzed to identify the meaning or significance in indicating how Catherine felt. Each emotion was regarded as a thematic category. To acknowledge the subjectivity associated with assigning meaning to emotion labels, common definitions for the terms were agreed between Catherine and Zembylas and member checking and triangulation of data sources were undertaken. Catherine's reactions to Zembylas' interpretation provided more insight and stimulated further reinterpretations and conversations about how she felt.

Fifteen emotion metaphors used by Catherine over the 3-year case study are reported, alongside an example of each metaphor and its frequency over 3 years. The metaphors capture aspects of emotion including intensity, control, positive–negative evaluation, difficulty, and harm. The emotions include frustration, excitement, joy, sadness, anger, love, happiness, pride, shame, and fear. Frustration as a natural force was the most frequently used metaphor (e.g., "I felt overwhelmed by frustration") with 48 occurrences, followed by Excitement as health (e.g., I feel great! I am so excited about all the ideas that you are coming up with …") with 32 occurrences. Happiness as a release of energy (e.g., I was flying high! I was very happy they [the children] got it!) and Excitement as an animal behavior (e.g., I was so excited, I jumped out of my skin! It was unbelievable how much imagination and creativity they [children] demonstrated" were the least frequently used metaphors with 4 occurrences each.

The findings reveal there is a disconnect between Catherine's views about the role of emotions in teaching science compared with those held by her colleagues. In addition, the emotional labor reported by Catherine is not readily recognized by her colleagues which Zembylas (2004a) attributes to the hidden views about how science should be taught and that is with an emphasis on specialized knowledge and professionalism. The study also found that there are both positive and negative aspects of emotional labor in science teaching in that it can lead to frustration, disappointment, or self-alienation but also increased satisfaction, self-esteem, and psychological wellbeing. Zembylas reports the importance of examining emotional meanings of metaphors to understanding a teacher's beliefs in a context. Moreover, he argues that "emotion metaphors affect a teacher's emotional behavior and experience because language influences how one acts and behaves" (Zembylas, 2004a, p. 317). The analysis demonstrated that Catherine sought out emotional labor as a source of satisfaction and reward, and in doing so, Zembylas suggests that Hochschild's (1983) conception of emotional labor, which informed the study, could be broadened to consider the positive aspects of this element of teachers' work. Finally, the study highlights the connection between professional discourses within a field of teaching practice and the display of emotions. What is viewed as the appropriate way to teach science in any given context shapes the emotional labor experienced by the teacher. Zembylas (2004a) suggests that containment of emotions and distancing may be two key strategies deployed by science teachers to remain consistent with the view of science that privileges reason over emotion.

Zembylas (2004b) again drew on data from the same 3-year ethnographical case study of Catherine, to examine how she used emotions to negotiate the *emotional tone* of her classroom and how she navigated positive and unpleasant emotions in science teaching.

Data analysis showed that Catherine's decisions drew on a constructionist view of emotions—that is, that the culture of the classroom is created through social interaction. As such, Zembylas (2004b) found that the emotional responses of both Catherine and her students played a role in their mutual construction of a supportive emotional tone in the classroom. Furthermore, emotions played a role in negotiating and re-negotiating the emotional rules that regulated classroom activity while learning science.

Two vignettes offered insights into the social construction of emotions in Catherine's science teaching. Vignette 1 focused on excitement and wonder (pleasant emotions) and Vignette 2 on frustration (unpleasant emotion). Through the data analysis, excitement and wonder were found to directly and indirectly influence Catherine's pedagogical decisions and actions. Teaching science in a manner that seemed intellectually and emotionally exciting led to emotional rewards for Catherine. In addition, Catherine's work demonstrated her understanding of the interrelationship between conceptual understanding and pleasant emotion. Vignette 2 represented the frustration, alienation, dismissal, and failure experienced by Catherine due to the misalignment between her personal teaching philosophy (student-centered, integrated units of work) and the school philosophy which was focused on student performance on standardized tests. Zembylas (2004b) connected these experiences to the demoralization of teachers which can lead to negative impact on self-confidence, self-esteem, and questioning their role within the profession. Here, Zembylas draws connections to shame and self-confidence being integral to identity-building and therefore self-esteem. He suggests that for elementary school teachers who may feel uncertain about their abilities to teach science, a supportive emotional culture is critical to maintain their sense of acceptance and avoiding marginalization from the system and profession. The paper concludes that, when theorizing emotion formed through social relationships, they may be seen as constitutive of the activity of teaching and, moreover, examinations of how teachers navigate school discourses are integral to theorizations of self-confidence, science teaching, and teachers' emotions.

An interesting outcome we identify across Zembylas's four articles focusing on Catherine is that he adopted a range of different emotion concepts in each article (Zembylas, 2001, 2002, 2004a, 2004b). The different outcomes about Catherine's work that were afforded by different theoretical and conceptual foci demonstrates the importance of adopting a variety of emotion theories in education and science education research. Each theory or concept can illuminate different dimensions of preservice or inservice teachers' work, which, when brought together, provides a more comprehensive picture of the role emotions play in science teaching.

Extending ideas on emotional labor we encountered in Zembylas (2004a), Bellocchi et al. (2018b) present sociological research on teacher emotion work in science and mathematics classrooms using narrative vignettes. The study aims to understand the emotional nature of teachers' work, and how it may interrelate with teacher attrition. Reflections of past school teaching experiences by Bellocchi and Mills (the first and second authors of the article respectively) form the data sources.

In the vignette drawn from Mills's experiences of teaching science in a primary school setting, she notes that "the emotional costs of teaching gradually seemed to become heavier" (p. 250) which she links to the tensions arising from navigating external measures of accountability and curriculum development which contributed to her view of the de-professionalization of teaching. She describes other pressures associated with building and maintaining relationships with parents and colleagues that required her to become resilient. She described disaffected students and the need to navigate what she observed as inequities experienced by students. She concludes by stating that by the end of her seventh year of teaching it was not her knowledge or skill that was limited but, rather, it was that her emotional resources were depleted to that points that she could no longer remain in the classroom. Through the analysis of Mills's vignette, the researchers identify significant implications for research, policy and for teacher education practice including the reminder that emotion is often regarded as irrelevant in the fields of science and mathematics, however teaching involves emotional work and as such, teaching science and mathematics presents the need to recognize the importance of an emotionally supportive organizational climate, at least within which this work can be undertaken. The researchers also note that the previous research has highlighted gendered differences in emotion work with women more likely to engage in 'loose emotion work', for instance, listening and counselling; whereas men are more likely to engage in 'tight emotion work' such as directing and controlling the emotions of others and that Bolton and Boyd's critique of the Hochschild's concept of emotional labor is that sometimes emotion work is an example of the best practices between colleagues. Here, the suggestion that collective emotion work is a consideration that could be highly significant to the retention of science teachers.

Vignette 2 focuses on Bellocchi's experiences when teaching mathematics in a Grade 8 classroom at a secondary school. In this narrative, Bellocchi experiences of the classroom lead him to state that the work he was engaged in—wrangling students shouting profanities during an altercation—did not feel like the work of a science-mathematics teacher. Over time, Bellocchi built relationships (a form of emotional work, also described by Mills) which developed his reputation within the school community. Those social bonds became a resource for him to draw upon to sustain his practice. In addition, Bellocchi also learnt to regulate his reactions to negative classroom experiences. In particular, he articulates three key stages of his own professional transformation in relation to the emotional work he was undertaking. In the first stage, Bellocchi reacted without mediating his emotions. The second stage saw him suppress the emotions he was experiencing. Then, in the third phase, he began to manage the expression of his emotion—first through facial expression, then with a calm voice stating the emotion, then issuing a consequence. The research team notes that emotional reflexivity is key to Bellocchi's transformation in that the interdependence of reason and emotion is revealed as central to the process of developing as a teacher.

When comparing the two vignettes, the researchers found that the capacity for a teacher to sustain emotional work is limited by their role identity. Teacher burnout occurs when science-teacher role identity is not congruent with their personal identity. By addressing emotional labor in the context of both role identity (i.e., teacher identity) and personal identity, Bellocchi et al.'s (2018b) study extends Zembylas' (2004a) work, which reported how the lack of recognition of emotional labor by Catherine's colleagues led to impacting her *teacher identity*. Furthermore, in addition to the limitations of the emotional labor construct reported by Zembylas, (2004a), the vignettes in Bellocchi et al.'s (2018b) study highlight limitations of Hochschild's emotional work concept, which does not adequately account for the social dimensions of emotion, instead focusing on the individual, cognitive dimensions of emotional work. Bellocchi et al. (2018b) also recognize *temporality* (conceptualized as a lack of time, or change over time, or intensification of work through time) as a factor that is linked to the emotional work of teachers. The paper concludes by suggesting that teacher education curriculum could include more explicit attention to reflective lessons, with suggestions on the effectiveness of emotion metaphors to facilitate such work. In addition, the authors suggest that emotional work undertaken as part of teacher education curriculum could be framed usefully by Billett's (2011) socio-personal conceptualization of work integrated learning as a productive way of attending to the conceptual, procedural, and dispositional knowledge needed to navigate the work of teacher in a professional setting.

9.3.2 The Role of Emotions in Ongoing Professional Learning

Professional learning of science teachers was the focus of three included studies. Drawing on CVT, Thomson and Turner (2019) noted that "teachers' emotions and values about their professional development (PD) experiences greatly influenced their learning of new knowledge and triggered positive changes in their classroom teaching" (p. 8). Emotions were identified by participants before and after participation in a 6-week professional learning experience focused on science methods and practice. Teachers were asked to identify the emotion they experienced before and after the professional learning program via a survey that identified 10 emotions (excited, anxious, inspired, disappointed, angry, overwhelmed, determined, reluctant, confident, content). Findings of note included gains in science knowledge made through the PD participation triggered pleasant emotions; feeling excited and inspired related to changes in teaching practice; both pleasant and unpleasant emotions triggered positive values and outcomes. For example, unpleasant emotions experienced at the beginning of the program served as motivation for learning content which, by the end of the program, triggered pleasant emotions and led to changes in teaching practice. The other key finding related to the authenticity of the learning experience whereby teachers worked with a scientist in a lab for an extended period. This attribution occurred despite the researchers' acknowledgement of the role that an authentic learning environment played in allowing "teachers to learn and process new information (i.e., science knowledge, procedures and relationships) in various ways and make more solid connections to their practices when applied in classroom teaching later" (p. 9). There is a missed opportunity here to explore the extent to which the environment worked as an antecedent to teacher emotions. Thomson and Turner's (2019) connections between emotion and science content knowledge are a feature of the second included study on science teacher emotion and PD.

In the second study of science teacher professional development, socio-emotional risk and comfort are the focus. Although the study involves interactions between an inservice and preservice physics teacher, we include it in this section on inservice teachers. Mahmood et al. (2024) report a case study of two Midwestern US science teachers who took part in a physics PD. The authors of the article delivered the four-day PD program, and the goals were to encourage teachers to embrace new curriculum and pedagogy and share resources. Audio-recordings captured interactions between a veteran teacher and a preservice teacher as they discussed physics examinations. Following narrative-style analysis, findings showed that interactional moves in the teacher-preservice teacher conversations can build comfort and reduce socio-emotional risks, which creates an environment where teachers feel safe to discuss physics content. Four specific interactional moves facilitated comfort including challenging the need to find one correct answer to physics problems, showing vulnerability, collaboration-centered problem solving, and storytelling. Emotional risk arose for the preservice teacher when questioning her knowledge of physics as the pair reviewed examination questions. The veteran teacher played a key role in initiating comfort-buildings interactional moves, including diverting from discussion of content that undermined the preservice teachers confidence, expressing her own vulnerability relating to physics content, and lastly through the exchange of personal stories that served to normalize beginning teachers' insecurities. One feature of the veteran teacher's moves was that they established trust with the preservice teacher, which is a topic at the center of González-Weil et al.'s (2025) study.

The third study of professional learning addressed the question of how trust can be fostered and sustained within professional learning communities (PLCs) in contexts where mistrust is deeply embedded in educational culture (González-Weil et al., 2025). The authors explored the development of trust in a Chilean PLC of science teachers over an 11-year longitudinal study conducted between 2010 and 2021. Although PLCs can foster productive spaces for teacher collaboration and reflection with a view to transform teaching practices and enhance student learning, in Chile, the development of PLCs is particularly challenging, as Latin American school systems and education models are based on mistrust, and competitive, individualistic school structures have historically hindered collaboration.

The PLC was comprised mostly of secondary biology, physics, and chemistry teachers, but it also included, for example, a primary teacher and mathematics teacher, as well as scientists who work in science teacher education. Data were collected across two stages of the PLC: Stage 1 (2010–2015) and Stage 2 (2018–2021). In the first stage, 12 transcripts of the PLC's 2-h working sessions (including two focus groups) were analyzed. In the second stage of the study, transcripts drawn from eight working sessions (wherein lesson stories were presented and socialized by different teachers in the group), four focus groups and four interviews with two teachers were analyzed. Thematic data analysis in the second stage included an in-depth exploration of the category of trust with a view to understand how it developed over time.

The study identified a progressive, layered development of trust in the PLC over time, from the initiation of the group during the first stage of the study, through to consolidation and transformation of the group during the second stage. Although the authors report a range of significant fundings about the ways in which trust was developed and maintained in the PLC, we address only the findings relate directly to the teachers' emotional experiences here.

In the second stage (2018–2021), a consolidated space of trust became apparent as the group met regularly. Teachers' free expression of experiences and emotions were critical in establishing a space of psychological safety, wherein teachers could share failures, vulnerabilities, and emotions without judgment such as feeling afraid about teaching hybrid classes during the pandemic, or happiness at sharing a positive lesson story when a teacher re-engaged with their students.

The study offers an account of how trust can be built and sustained in a deeply mistrustful educational system. The findings of the study illuminate the development of trust as a critical element of the PLC over time, facilitating the development of a highly valued space for professional learning wherein teachers felt comfortable freely sharing their emotions. In this way, the PLC became a safe space for teachers' emotional support. This finding supports the key idea evident across the three studies, which show that PD can provide confidence and build teachers' science knowledge through the emotional support provided by the professional learning communities.

9.3.3 Antecedents to Teacher Emotions

Antecedents are the events or stimulus that triggers a behavior (environmental—a noise, or an object; social/interpersonal—interaction with someone; internal (a feeling or emotion). In Membiela et al.'s study (2023), antecedents of the emotions most frequently experienced by 430 Chilean inservice teachers when teaching science was the focus. A range of factors identified from previous research were selected as relevant antecedents of teacher emotions including: motivation towards teaching career; attitude towards science teaching; teacher self-efficacy; and satisfaction with teaching career choice. as relevant antecedents of teacher emotions. Emotions are an affective arousal mechanism associated with tasks like studying or results such as success and failure. The study found that variables associated with motivations towards teaching career are strong antecedents of teacher emotions. Teacher self-efficacy is also recognized as an antecedent of teacher emotions. In terms of the emotion of enjoyment, *relevance of science teaching* and *satisfaction with teaching career choice* were found to be relevant antecedents. The authors suggest that based on their findings, initiatives aimed at improving the intrinsic value of the teaching career, the value of working with students, and teacher self-efficacy are critical to improving teacher satisfaction and triggering pleasant emotions.

9.3.4 The Influence of Emotions on Teachers' Pedagogical Decision-Making

The underexplored role of teachers' emotions on science teaching in early childhood education (ECE) is the focus of Praderio Gaias et al.'s (2024) study. Acknowledging the importance teachers' emotions and early, personal science education experiences play in their teaching practices, the authors investigate how ECE teachers' emotional experiences—both past and present—influence their decisions when planning and implementing science lessons.

The researchers employed a qualitative multiple case study design focusing on three experienced female ECE teachers from an ECE center servicing 53 children aged 3–5 years in La Plata, Argentina. The teachers elected to participate in the study given their own reported difficulties and lack of self-efficacy in teaching science. Data collection spanned three phases: pre-active (planning), interactive (classroom implementation), and post-active (reflection). Data were collected through a survey on emotions in science teaching; document analysis of lesson plans; video-stimulated semi-structured teacher interviews; and classroom observations and video recordings. An observation guide was developed to record the teachers' emotions, as indicated by their facial expressions, tone of voice, and body language and gestures.

The first part of the emotions survey was comprised of closed-ended and multiple-choice questions, adapted from Borrachero et al. (2011), that examine teachers' emotions, states of mind, feelings, and attitudes associated with their experiences of learning science from their own primary schooling through to their initial teacher education. The teachers were also asked to identify at which educational level and in which science discipline (biology, chemistry, or physics) they experienced specific emotions. The second part of the survey comprised of an adaptation of Bandura's (1977) perceived self-efficacy scale. It asked teachers to identify the emotions they associated with teaching science content in ECE, and how prepared they felt to perform teaching-related tasks (generally and in science, specifically).

The study's reveals a complex and dynamic relationship between teachers' emotions, their science teaching practices, and classroom dynamics, as presented across three key dimensions:

1. ***Emotions associated with their previous experiences of learning science:*** Teachers recalled pleasant emotions such as curiosity and pleasure mainly in relation to biology during primary school. In contrast, unpleasant emotions (e.g., frustration and boredom) were associated with chemistry and physics, particularly during secondary education and their initial teacher education. It was found that these past emotional experiences were significant because they had a strong impact on how the teachers approached their current science teaching practices.

2. ***Current Emotions associated with science teaching:*** Despite mixed learning experiences in science, all participants expressed predominantly pleasant emotions when teaching science, such as satisfaction, joy, and curiosity. These emotions were linked to high self-efficacy scores, reflecting strong confidence in their ability to teach science effectively. The study also found that pleasant emotions motivated teachers to enact more interactive learning opportunities, while unpleasant emotions restricted teachers' pedagogical choices and led them to avoid more complex science topics.
3. ***Emotional influence on pedagogical decision-making during lesson planning and enactment:*** In the planning phase, emotional memories and anticipated emotional outcomes (i.e., affective forecasting) influenced teachers' topic selection and teaching strategies. Teachers often chose science topics that evoked pleasant emotions or past success (e.g., human biology, dinosaurs). During the interactive phase, emotional responses to classroom dynamics (particularly student behavior) prompted adaptations to teaching approaches. Unpleasant emotions, such as frustration and concern, often led to shortened lessons or reduced student engagement. In the post-active reflection, teachers acknowledged how emotional experiences shaped both instructional decisions and classroom climate, often modifying future strategies accordingly.

The study' significance lies in its key finding that both experienced and anticipated emotions are significant in shaping teachers' pedagogical decision making when planning and enacting science lessons. Additionally, although pleasant emotional experiences foster teacher engagement, persistence, and innovation, unpleasant emotions can constrain pedagogical decision-making. With this noted, the study did not explore how specific anticipated emotions affect various aspects of science planning and teaching, nor did it examine the long-term impact of teachers' anticipated emotions on their pedagogical practices. These angles are worthy of further research.

The authors posit that teachers' emotional development should be an important part of initial teacher education and professional development programs, advocating for approaches that foster pleasant emotional engagement with science, enhance teachers' self-efficacy and emotional awareness (cf. Membiela et al. 2023), and equip teachers with emotion regulation strategies to help manage their unpleasant emotions, to enhance their science teaching.

9.3.5 Emotions and Career Changing Scientists

Career changing scientists or second career scientists are preservice or inservice science teachers with university qualifications in science, who subsequently study teacher education (Bullough, 2009; Ritchie et al., 2011). These educators are at times recent science graduates who switch to education and become teachers, or people who have held careers in science for varying lengths of time before making the switch.

It is common for these science teachers to have very deep knowledge of science, albeit often in one specific discipline, and experience with applications of science in research or applied endeavors that confers some advantages in their teaching. However, lack of familiarity with school science classrooms, limited experience with school and young people in learning contexts, the short education courses they often complete in recognition of prior studies, and the rapid timeframes in which the transition to becoming teachers occurs can present situations that are highly emotional and may be experienced for the first time. Expectations that career change scientists have for teaching collide with school and classroom realities during practicum and work. The identification of one empirical journal article (i.e., Ritchie et al., 2011) and one case in a book chapter (i.e., Bullough, 2009) suggests that we need to learn much more about the experiences of career-changing scientists' emotional experiences in preservice and inservice science teaching.

9.3.6 Emotions and Managing Science Classrooms

The studies included in this chapter also draw a link between teacher emotions and the management of science classrooms. Drawing on CVT, Uzuntiryaki-Kondakci et al. (2022) noted that teachers' capacity for emotional regulation is key to avoiding burnout. For the teachers in that study, the most frequently reported unpleasant emotions are expressed in relation to classroom management, and these are anger and worry. Uzuntiryaki-Kondakci et al. (2022) also noted emerging evidence around links between emotions and pedagogical content knowledge (PCK). Teachers who consider student misconceptions and difficulties–as an indicator of PCK—experienced more pleasant emotions. This element of PCK contributes to the positive affect of the classroom. In addition, teachers with mastery orientation towards goals, and higher self-efficacy beliefs experience pleasant emotions more frequently. The study found that when teachers enact student centered pedagogies, more pleasant emotions are experienced in the classroom, with the two most reported emotions being enjoyment and contentment. However, in a much earlier study, dos Santos and Mortimer (2003) noted that strong PCK alone is not enough to maintain positive affect in the classroom. Instead, teachers needed to monitor continuously their development of, and responses to, emotions, feelings, and affect. dos Santos and Mortimer (2003) suggest knowledge about teachers' emotional work should be considered critical to the development of a teacher's PCK.

Strategies to support teachers' development emotional regulation are suggested by Brick et al. (2021). This study is set in Liberia and focuses on using neuroscience to develop teacher capacity to understand and support students' social, emotional, and behavioral issues with the unanticipated outcome being that those teachers, themselves, also developed knowledge for supporting their own emotional regulation. The context is of significance to this work as it acknowledges that Liberian teachers, themselves, have unmet emotional needs from traumas arising from civil war and pandemics. The authors note that the capacity for education to support and prepare students and teachers for and through schooling was seriously eroded during these times of social disruption and unrest. As such, professional learning opportunities that develop teachers' knowledge of the impact of trauma on brain development was found to provide emotional support for teachers and improves their skills to handle the social, emotional, and intellectual needs of their students. The teachers who participated in the study identified that they were able to adapt their approach to classroom management once they understood the experiences of learning and teaching in science through a trauma-informed lens. While this study is set in a Liberian context, there are broad implications for the work of science teachers who serve schools and communities that have experienced, or continue to experience, trauma. Acknowledging the emotional work required from teachers to develop and sustain a positive classroom environment validates their experiences as educators and provides strategies to support the development of these capacities with a view to avoiding emotional burnout.

9.3.7 Emotional Support for Fostering Inservice Teachers' Conceptions of Creativity

Fostering school students' creativity has been central impetus in policy and curriculum (Hong & Kang, 2010). Enthusiasm for consideration of creativity in schools seems related to notions of developing students' personal fulfilment and their future economic potential. Beginning with the assumption that cultural values and contexts may shape a teacher's conceptions of creativity, Hong and Kang (2010) compared South Korean and US inservice science teachers' conceptions.

One difference between South Korean and US science teachers' conceptions of creativity relates to emotional support. Based on survey responses relating to specific pedagogies for fostering creativity in the classroom, South Korean teachers emphasized provision of opportunities for science students to think through open-ended problems and to engage in discussion through cooperative learning. Providing encouragement, a classroom environment factor, is interpreted by Hong and Kang as a form of emotional support. The different focus on cognitive aspects by South Korean teachers and emotional support from the learning environment is attributed to differences in curricular emphases. The National Standards in the US draw attention to learning environments, whereas the South Korean curriculum focuses on divergent thinking.

A socio-cultural explanation is offered to account for differences in pedagogical foci between teachers from the two countries. Ethicality was a focus for South Korean teachers, and this is attributed to an Asian cultural emphasis on group goals (Hong & Kang, 2010). An individualistic approach associated with self-aggrandizement is attributed to Western culture and used to explain US science teachers' greater emphasis on the learning environment and emotional support. The somewhat contradictory nature of this claim is not addressed, and it is difficult from this study to draw any solid conclusions about what role emotional support may play in supporting science students' creativity. A more interesting observation offered in Hong and Kang's findings is what role curricular emphases may have on science teachers' awareness of emotions and emotional support as important factors shaping learning. One of the key blank spots for science education research on emotion is the connection between levels of social reality and emotions. Studies tend to operate predominantly at the micro-level of the classroom or individual teacher. The pattern observed by Hong and Kang between curricular emphases (a meso-level social structure) and the teachers' perspectives reflecting those emphases suggests that connections between micro-level classroom practices and meso-level structures such as curriculum can be studied. A step beyond this would be to consider social institutions (e.g., an education system), the macro-level, and how emotions operate across all three levels.

9.3.8 Emotions and Bilingual Science Education

In a study set in Slovak schools Kordíková and Brestenská (2022) examine the bilingual teaching of science. Draws on theory of semantic differential, a rating scale used to gauge respondents' subjective attitudes (the affective and cognitive components of a response) to selected concepts and is regarded as a more reliable and valid approach than scaled questionnaires because it gives greater data consistency in quantitative factor. The scale has 20 bipolar adjectives. Of the 20 pairs of adjectives, 8 were related to emotional satisfaction of the teachers: exciting-boring; friendly-unfriendly; beneficial-meaningless; making joy-driving fear; valuable-worthless; interesting-dull; pleasant- unpleasant; good-bad. The findings showed that inservice teachers experiences slightly higher emotional satisfaction when teaching science in English than the preservice teachers who were also participants in this study. The authors suggest that this may be attributed to the experience of teaching in a foreign language (in this case, English) provided new ideas and possibilities for rejuvenating their teaching practice, however this suggestion is not grounded in the findings of the research. In comparison, preservice teachers experienced lower emotional satisfaction when teaching science in English as compared to inservice teachers, even though they considered the teaching of science in both languages to be beneficial and valuable. The study makes suggestions that this could be attributed to a lack of experience in bilingual teaching, and no possibility to prepare for teaching in foreign languages prior to entering a school setting. From here, we move to examine the links between emotions and the work of preservice teachers.

9.4 Empirical Outcomes on Emotion in Relation to the Work of Preservice Science Teachers

Emotion plays a role as preservice science teachers work to develop their capacity to facilitate science learning experiences. These can be emotions that preservice science teachers navigate as well as the emotions of their students and colleagues.

9.4.1 *Emotions and Learning to Teach Science*

Zembylas and Buhlman Barker (2002) took an epistemological stance that learning how to teach science is "a deeply emotional activity in which the individual concerned has to deal with his or her prior emotions and attitudes in the light of their past and present experiences" (p. 346). This paper examined the impact of creating time and space within a preservice teacher education program for preservice teachers to engage with their attitudes and beliefs intellectually and emotionally about learning to teach science. The authors developed two pedagogical innovations; individual spaces which include emotional and intellectual elements as well as material affordances in an effort to foster emotional comfort while they engage in reflective practice, alongside community conversations which are conceptualized as 'discourse community' within which preservice teachers can broaden their understanding of the attitudinal transformations that underpin their developing knowledge of science and science teaching. The work utilizes an in-depth case study to examine the work of two preservice teachers—Jenny and Scott—as they engage in their teaching methods course which incorporate the analytical tools of individual spaces and community conversations as a means of facilitating attitudinal transformation. Jenny's transformation is characterized by the authors as moving from positive feelings to even more positive ones. Scott, on the other hand, began by expressing mostly unpleasant emotions (frustration, anxiety uncertainty) and negative attitudes towards science based on narrow, empiricist experiences of science from his time at school. When reflecting on his transformation he attributed his time in the methods class as an opportunity to consider the origin of his attitudes towards science and to create new ideas about how science could look and feel for students in his own future classroom. The authors suggest that the stories of the preservice teachers' experiences point to what should be a critical goal of science teacher education, and that is help preservice teachers to develop both an emotional and an intellectual understanding of their work—in relation to both their current and future practice— through the analysis of their emotions and attitudes about science and science teaching.

Emotional or affective habits and their role in situated emotional or affective experiences are the focus of Ma's (2018) exploratory qualitative study in which she interviewed 9 preservice elementary teachers completing a teacher education course in Australia. Using semi-structured interviews, Ma (2018) asks the elementary preservice teachers to recall specific examples of their secondary school science learning experiences. Through thematic analysis, the responses are grouped into six categories; three reflecting pleasant emotional experiences and three reflecting negative experiences. The analytic framework focuses on relationships between situated affective experiences (i.e., emotions experienced in a specific situation) and habituated affective attributes (i.e., enduring emotional predispositions). The classroom learning environment is considered an important external factor that shapes both types of emotional experience. The two are related in that predispositions are considered to shape or impact situated emotional experiences. Preservice teachers who feel highly self-efficacious most likely also experience fewer unpleasant emotions during teaching.

From analysis of interview data, interesting interactions between habituated affects and situated affects are identified. In the positive experiences, habitual emotional disposition can overcome the effects of unsatisfactory learning environments leading to positive situated emotions. In contrast, a well-designed classroom learning environment that exceeds student expectations has the capacity to overcome negative habituated emotions towards science. Lastly, there were examples where there was direct correspondence between the habituated emotions and the situated emotions, such that positive situated emotions are elicited by the environment when the student holds pleasant emotional dispositions. This category was associated with cognitive aspects of science classes and peer relationships, where these aspects corresponded with student expectations from their habituated emotions.

Negative situated emotions also, predictably, resulted from the nature of the classroom learning environments. In the first category, the elementary preservice teachers recall cases where the delivery of classroom instruction conflicts with their beliefs or where they have personal conflicts with the teacher. Although these past science students may have held either positive or negative habituated affect towards science, in the situations described, the lesson delivery and personal clashes with teachers led to negative situated affective experiences. The second category of negative situated emotions relates to experiences of disappointment and dissatisfaction when students enter science class with high expectations that are not met by the learning environment. Aspects of the learning environment that stymied positive anticipatory emotions included not understanding lesson content in the way of textbook information, repetition and boredom from reading and answering questions, and the failure of classroom lesson to match the interest and engagement generated by wildlife documentaries and other media sources.

In the third negative situated affect category, the learning environment is considered irresponsive to the habituated affect of students. In this case, positive habituated affect such as interest, is not activated, or negative habituated affect such as anxiety about science is not addressed. The key difference here with the second negative category is that students did not enter the science lesson with high or positive expectations. Due to the inability of the environment to produce positive experiences and curb their negative habituated affect, negative situated emotions results. Overall, Ma (2018) suggests that the findings of this study show that neither habitual affective attributes (HAAs) nor classroom learning environment (CLE) alone can accurately predict positive situational affective experiences (SAEs). Whether or not a student will have a positive science learning experience results from the interplay between HAA as a prior factor, and CLE as a mediator of the experience.

What is interesting and unique about Ma's (2018) study is the comparison of habituated emotion to situated emotion. There are several limitations to the study, however, which challenge the validity of some of the claims. The first is that with the assertions made in each category Ma identifies, there is an inherent causal association implied that is not justified or supported based on the study design. Secondly, Ma's study presumes a classical directionality relating to emotional experiences, where the environment is the stimulus and the emotion is the internal response of the organism, in this case the elementary preservice teacher as a past secondary science student. Considering that theories of emotion long ago moved away from behaviorist tendencies when thinking about emotions, it is somewhat surprising to find this logic reappearing in Ma's study. Some of the challenges this creates for interpretation of data in Ma's study is that the preservice elementary teacher's emotions are seen to result predominantly from the environment, where they are positioned by the theoretical approach as passive recipients. That is, the classroom environment does things to these past science students, and the students form habits or situated emotions in response.

This somewhat simplistic view is problematic because it is not addressed through the habituated emotion construct which also assumes the same behaviorist logic when explaining how the habitual tendencies were formed in the first place. Although the analyses are interesting, contemporary emotion theories could be used to shed new light on the interplay between habituated emotions and situated emotions, which presents an interesting avenue for further study. Another theoretical aspect that requires attention is the adoption of models of interest, widely considered an affect, which differentiate between habituated and situated affect, to conceptualize emotions. Ma acknowledges that emotions are predominantly considered to be situated short duration phenomena, but the relationship assumed theoretical between situated emotion and habituated affect or emotion is tenuous and requires greater theorization and empirical exploration. Much of the analysis based on habituated affect can be understood from expectations state theories of emotion, so it is not clear what difference, if any, exists between the approach and assumptions Ma has taken and existing expectations states theories of emotion (cf. Turner, 2007). There is some conflation between emotion and affect that limits what understandings about emotions are central and relevant to the learning environment. Greater focus on emotions would enhance our understandings of specific emotions and their role in preservice elementary teachers' dispositions towards science and science teaching.

A third limitation is what value and rationale can be upheld for asking preservice elementary teachers to recall secondary school experiences. To the extent that this answers questions about high school students' recollections of emotion and affect in science, I would agree. However, one must draw a long bow if one wishes to infer any meaning from these past experiences in understanding preservice elementary teachers' experiences of their university education classes, future teaching, or dispositions towards teaching science. Ma's exploratory study does establish preservice elementary teachers' perceptions of their past school science experiences, and future studies could establish such historical life experiences as they track preservice and inservice elementary science teachers longitudinally. I agree with Ma that ascertaining what effect, if any, historical emotional experiences have on preservice education experiences and inservice teaching experiences is a valuable area of future inquiry.

In the study by Wicaksono and Korom (2023), data were collected from 296 students enrolled in three universities in Central Java, Indonesia. 56.08% of participants were studying to become science teachers, and 43.92% were majoring in Engineering. Participants completed the Attitude Towards Science Questionnaire (ATSQ) (van Aalderen-Smeets et al., 2012), adjusted by the authors for the Indonesian context. Of the dimensions of attitudes towards science, the second dimension—affect—is related to the positive and unpleasant emotions preservice teachers report about science. Pleasant emotions are manifested by the enjoyment of science and confidence in science, while unpleasant emotions are represented by anxiety and difficulty with science. Correlations between attitude variables (enjoyment and confidence; anxiety and difficulty; participation in science learning and activity; value of science) were calculated and no significant difference was found in attitude science on any of the dimensions between science teacher candidates and engineering students. Both science preservice teachers and engineering students demonstrate strong positive correlation between enjoyment and willingness to participate in science learning and activities as well as their perception of and value of science. However, correlation patterns did differ in relation to the variable anxiety and difficulty- for the Engineering students, there was a weak significant positive correlation between anxiety and all of the other variables suggesting that situational factors are at play in the anxiety experienced by the Engineering students.

In the study by Erdemir and Kandil İngeç (2023), which focused on assessing emotional empathy employed by preservice teachers as they worked to develop conceptual understanding of the concept of energy through the use of metaphor. Preservice teachers were found to have high levels of empathic tendency and the scores in the "you level" of the empathic tendency scale were prominent. This results in preservice teachers exhibiting behaviours such as supporting others, addressing problems, repetition, and understanding deep feelings. The authors interpreted these findings as evidence of the significance of empathy as an emotion in the work of math and science preservice teachers as they looked at events and situations from the other person's point of view.

9.4.2 Emotion and Learning to Teach Science Across Disciplines

In a body of work conducted in Spain, four papers examine interactions between emotions and learning to teach science across disciplines. Results reported across these papers suggest that it is important to avoid the homogenization of the emotion involved in learning to teach 'science' as different preservice teachers will experience different emotions in relation to different disciplines such as biology, chemistry, and physics. Brígido et al. (2010) study of 63 primary education students enrolled at University of Extremadura, Spain, completed a questionnaire, organized in terms of tables of emotions (Brígido et al., 2010). The study found a strong correlation between the recall of emotions of learning sciences as a secondary school student and emotions of teaching science. Strong unpleasant emotions in learning and teaching physics/chemistry with unpleasant emotions such as nervousness, anxiety, tension, worry, and despair reported. In contrast, strong pleasant emotions in learning and teaching nature sciences. Overall, the study reports that preservice teachers' memories of emotions in learning science at school are more negative than those experienced when teaching science. By gender, men reported more pleasant emotions for teaching science than women. The authors also recognize several limitations of this study including the small sample size and the lack of nuance of emotion that can be extracted from the questionnaire responses and suggest that follow up interviews could address this limitation. The implications for this study are framed in relation to recognizing how emotions differentially affect teaching and learning across science subjects and that, therefore, it may be necessary to implement programs within preservice teacher education to support the development of emotional competence of preservice teachers embarking on a career teaching science.

Also at the University of Extramadura, Brígido et al. (2013b) conducted a questionnaire with 188 primary education students with no prior experience as teachers. 120 of these participants were women (63.8%) and 68 men (36.2%). The authors used descriptive statistics and correlation to identify possible relationships between self-efficacy and the emotions that arise in response to the expectation of future primary school science teaching. The results show that these preservice primary teachers express different emotions according to whether they are considering nature sciences, or what the authors refer to as the 'hard sciences': physics and chemistry. When considering the nature sciences, preservice teachers expressed predominantly pleasant emotions. In relation to physics and chemistry, emotions reported are predominantly negative. Drawing on theory from Bandura (1977), the authors connect these reported emotions to preservice teachers' self-efficacy. Their findings suggest that unpleasant emotions towards teaching physics and chemistry are significantly related to self-efficacy. Moreover, self-efficacy has little connection to the predominantly pleasant emotions preservice teachers express about their future teaching of the nature sciences. Limitations of the study again are grounded in the limitations of nuanced reports of emotion possible through the survey, but interviews are not used to supplement this work, despite have recognized this limitation with the cohort of 63 preservice teachers studied in 2010. The implications are framed in relation to preservice teacher education look to improve preservice teachers' self-efficacy and the emotions about physics or chemistry.

Brígido et al. (2013a) expanded the sample size of the previous study (i.e., Brígido et al., 2013b) by administering a questionnaire consisting of 27 positive and unpleasant emotion labels to 315 primary preservice teachers studying at three universities across Spain. 223 of the participants were women (70.8%) and 92 men

(29.2%). The study examined the emotions that preservice primary school teachers could recall about their learning of science in primary school, and biology/geology and physics/chemistry in secondary school. The study also asked the preservice teachers to state the emotions that they felt about teaching those content areas in the future. Finally, the results were examined for patterns related to gender and to the area of content specialization that each participant studied in secondary school. The results showed that, overall, gender did not influence recalled emotions and that the preservice teachers have mostly positive recall of their experience of learning science in primary school. However, the cohort of preservice teachers who had a science specialization in secondary school (22.2%) recalled mostly pleasant emotions associated with learning science than the preservice teachers without a science specialization. Of the preservice teachers with a science specialization in secondary school, mostly pleasant emotions were recalled about learning Biology/Geology and mostly unpleasant emotions about learning Physics/Chemistry. There was some correlation between the emotions remembered when learning these subjects, and the emotions that preservice teachers anticipate they will experience when teaching these subjects in the future. The implications for this study are framed in relation to the need to include emotional intervention programs to support preservice teachers to regulate their own emotions as well as those of their students as they navigate possible emotional vulnerabilities in relation to teaching and learning in both Physics and Chemistry.

The fourth paper from the previous group of collaborators (Borrachero et al. 2014) focuses on 178 prospective secondary school science teachers which includes a sub- sample of 66 Science and engineering graduates, studying at the University of Extremadura, Spain. Along with the cohort under focus, the goals of the paper also change focus when compared with those of Brígido et al. (2013a), above. In this study, the researchers move away from an examination of the recall of emotion. Instead, the prior experience with studying science/engineering at the undergraduate level, along with gender, become factors of interest. The data analysis determined the emotions experienced by preservice secondary school teachers when teaching biology/geology and physics/chemistry and compared those emotions between the two cohorts: (1) those who hold prior Science/Engineering qualification (i.e. the sub-sample of 66 preservice teachers), to (2) those without a prior science qualification. In addition, the study aimed to identify any significant relationships (using chi-squared test with 95% confidence intervals) between gender and the emotions experienced by preservice teachers in the two cohorts when teaching biology/geology and physics/chemistry in secondary school. The questionnaire implemented varied from previous related studies. This version included 12 positive and 12 unpleasant emotions chosen from the same classifications utilized in other articles included in this collection (e.g., Brígido et al., 2009, 2010).

Overall, all preservice teachers expressed more pleasant emotions about teaching Biology (fun, 39.7%; tranquility, 37.6%; sympathy, 34.8%; and, motivation, 33.7%). This pattern was held for the science graduate cohort, however higher percentages of pleasant emotions were reported (fun, 43.9%; sympathy, 43.9%; confidence, 37.9%). Significant relationships were found in the pleasant emotion of confidence ($p =$

0.019) and the unpleasant emotions of hate ($p = 0.013$) and sadness ($p = 0.044$). However, the only statistically significant relationship between gender and emotion within the science graduate cohort was the unpleasant emotion of nervousness ($p = 0.046$).

Similarly, overall pleasant emotions were reported when teaching Geology (motivation, 22.5%; sympathy, 20.2%; attraction and confidence, 19.1% each). As was the case in relation to Biology, the science graduate cohort exhibited slightly higher percentages of pleasant emotions (motivation, 33.3%; fun, 28.8%; confidence, 27.3%). The chi-squared test showed significant relationships between emotion and gender when teaching Geology: attraction ($p = 0.023$), confidence ($p = 0.007$), joy ($p = 0.008$), satisfaction ($p = 0.008$), anxiety (p = 0.041) and fear (p = 0.048). The relationships of statistical significance within the science graduate cohort were confidence ($p = 0.027$), enthusiasm ($p = 0.030$) and stress ($p = 0.032$). Men reported higher percentages of pleasant emotions, while women report higher percentages of unpleasant emotions in relation to teaching Geology.

The teaching of Physics elicited mostly unpleasant emotions, overall (stress, 30.9%; nervousness, 29.2%; fear, 28.7%; concern, 24.7%; and, anxiety, 23.0%). However, these patterns varied when examining the cohort of science graduates. For these preservice teachers, the highest percentages corresponded to pleasant emotions (motivation, 33.3%; fun, 28.8%; confidence, 27.3%). Across both cohorts, the chi-squared test showed there to be significant relationships between gender and the emotions of confidence ($p = 0.000$), enthusiasm ($p = .0009$), fun ($p = 0.006$), gratification ($p = 0.042$), joy ($p = 0.036$), pleasure ($p = 0.036$), satisfaction ($p = 0.002$), tranquility ($p = 0.032$) and stress ($p = 0.046$). When examining the cohort of science graduates, the statistically significant relationships between gender and emotion were for confidence ($p = 0.027$), enthusiasm ($p = 0.030$) and stress ($p = 0.032$). As was also the case for geology, men reported higher percentages of pleasant emotions, while women report higher percentages of unpleasant emotion in relation to teaching Physics.

In relation to the teaching of Chemistry, unpleasant emotions were most commonly reported, overall (nervousness, 27.5%; concern, 24.7%; stress, 24.7%; fear, 24.2%; and anxiety, 20.2%). However, for the science graduate cohort, pleasant emotions were more frequently experienced (fun, 36.4%; satisfaction, 30.3% and confidence, 30.3%). The chi- squared test showed there to be significant relationships between gender and the emotions of confidence ($p = 0.017$), enthusiasm ($p = 0.042$), fun ($p = 0.024$), joy ($p = 0.011$), pride ($p = 0.048$), satisfaction ($p = 0\ 000$), tranquility ($p = 0.042$) and pessimism ($p = 0\ 046$), with some of these relationships also significant in the science graduate cohort: fun ($p = 0.041$), satisfaction ($p = 0.007$), tranquility ($p = .049$) and pessimism ($p = 0.020$). Again, as was the case for both geology and physics, men expressed mostly pleasant emotions about teaching Chemistry, while women reported mostly unpleasant emotions. The findings are consistent with the other papers presented by researchers from this collective. They highlight the potential for teachers of physics and chemistry to experience unpleasant emotions such as fear and anxiety because of their work in secondary science classrooms. The authors note that women were more likely to report unpleasant emotions than men, regardless of whether the preservice teacher is a science graduate. Consequently, the authors call for preservice teacher education that includes emotional education.

9.4.3 Emotion, Identity, and Marginalization in Relation to the Work of Preservice Teachers

Science teachers from marginalized backgrounds form a unique focus and present interesting outcomes relating to emotions in teaching science. Emotional experiences of marginalization due to race are topics addressed by Moore Mensah (2019) and Rivera Maulucci (2011) with varying degrees of focus. In Moore Mensah's study, emotions arise through the narrative inquiry of one Black elementary science teacher's journey to becoming a teacher. In contrast, Rivera Maulucci's study offers an intentional focus on analysis of emotions by drawing on theories that inform interpretation of one Dominican elementary teacher who implements dual-language instruction in science in a US classroom. Both studies adopt narrative inquiry and critical theories as lenses for making sense of the experiences of these teachers who experience otherness both in their own schooling experiences and later as they become science teachers.

A common outcome of both narratives is that each teacher experiences obstacles early in her life and schooling due to her differences from the mainstream whether they are English speakers in Elena's case (in Rivera Maulucci, 2011) or White in Michele's case (in Moore Mensah, 2019). For Elena, the prospect of teaching science in Spanish brought emotions of fear and nervousness even though she is fluent in the language. For Michele, repeated experiences of feeling marginalized in a teacher education class where she felt othered and where Whiteness seemed the norm, she felt isolation, alienated, discouraged, and emotionally stressed. Experiences with early courses led to repeated experiences of not fitting in and being able to form a teacher identity that was commensurate with her other identities as a Black woman. Her emotions drove her to consider leaving the course, until she encounters a Black female professor in her science methods course. With the ability to identify with and find a mentor in this professor, Michele's story is turned around, and she proceeds with confidence into her practicums, even challenging the status quo when she is told to change a lesson she had prepared. This final course experience helped Michele to overcome deep seated identity issues beginning in her childhood and which she experiences time again throughout her early course experiences. The capacity to identify with another woman of color became the vehicle through which Michele found her voice and identity in science teaching and pursued the right school for her needs and position as an elementary science teacher.

Elena's emotional transformation from fear and nervousness to relief came with the realization that most of her students were not native Spanish speakers either and that she could use English for instruction, not just Spanish.

Both stories speak of the emotional challenges that being different from the mainstream can present teachers and how it influences the formation of an identity. In both cases, transformative situations bring about changes from unpleasant emotions to pleasant ones and these changes are associated with perseverance with the profession. Given that only two studies were identified that focus on the emotional experiences of marginalization, more work could be done to create deeper understandings and tease out the nuances of the emotions experienced by teachers who do not reflect the norm in schools or teacher education classes. Moore Mensah's (2019) observation on this is astute in identifying that most courses that consider teaching for the inclusion of diverse students tend to begin with the assumption that the teacher is White, and students who are different from the teacher require strategies for inclusion. From the lens of intersectionality adopted in her study, Moore Mensah considers the need that, in the case of her work, a Black preservice candidate may need to consider when preparing to be an inclusive teacher. As neither study addressed emotions in the research questions nor aims, a more direct exploration of emotions could yield further insights into studies of this kind.

In a second article by Rivera Maulucci, a case study of Nicole, a preservice teacher who identifies as African American, Caribbean, and who is enrolled in a methods course taught by the author, is presented (Rivera Maulucci, 2013). Through the case, the development of Nicole's social justice stance is examined with a focus on the role that emotions play directing her personal identity development as well as how her emotions can become strategies for her actions. This work draws on the theory of positional identity and the notion of the performative nature of teacher emotions to recognize that "the ways in which teachers understand, experience, perform, and talk about emotions are highly related to their sense of identity". This study explores emotions and positional identity at the intersection of race, class and gender and acknowledges the emotional ambivalence and emotional navigation that are integral to Nicole's experience of becoming a social justice Chemistry teacher. Through the analysis of Nicoles's case, both positive and unpleasant emotions were important to her transformation and resistance to ideas she met in her program of studies as she develops her identities of teacher, science teacher and social justice educator. This idea is linked to the work of Moore (2008) that recognizes the positional identity of preservice teachers with respect to both social justice and science, are neither static nor neutral. As a preservice teachers' goals and perspectives develop, so too does their positional identity, which will generate emotional ambivalence; that is to say that they will experience both strong positive and strong unpleasant emotions as they navigate the process of becoming a teacher. This was the experience of Nicole, and her emotions became resources for her. She was able to use them to make sense of her experiences as well as to guide change. The interpretation of Nicole's experiences is underpinned by the premise that emotions are not separate from cognition and, as such, emotions play a role in teachers' navigation of social justice issues. Firstly, through the identification and framing of teaching dilemmas and problems. Emotions are also central to teachers' reflection on such issues. During the reflection they may experience emotional ambivalence, evaluating and sense- making, and decisions concerning the positioning of the teacher themselves in relation to the issue. Finally, emotions—as mediated through emotional resistance and critical emotional praxis—are central to teachers' actions in relation to the social justice issue. The merit of critical emotional praxis as an approach that should be incorporated into teacher education is made clear through this research, particularly for preservice teachers whose own histories can inform their lives as teachers.

The notion that identity is also central to work by Ritchie et al. (2007) as they examine the story of a career-changing scientist, Tanya, navigating the process of becoming a science teacher. In this study, identity is viewed as discursively mediated through cognitive and emotional practices that arise between an individual and others sharing their cultural background. "Becoming a teacher then is a dynamic process of intersubjective discourses, experiences and emotion" (p. 229) as preservice teachers construct a hybrid identity (Bhabha, 1994) that bridges identities past, present and future. Using narrative research, the role of emotions experienced by Tanya during her transition becomes clear. Stories of success were key to eliciting pleasant emotions. Navigating the unpleasant emotions of her family and former colleagues as she decided to transition to a career in teaching also required emotional work in preventing the unpleasant emotions of others from undermining her own confidence in her decision to become a teacher might be prepared better for a diasporic life and identities through a program of storying their practices in varied contexts for different audiences. The implications of these findings again point to the inclusion of identity work in teacher education programs, particularly those programs that are developed to encourage career-changing professionals to become classroom teachers.

9.5 Conclusion

The most numerous studies ($n = 29$) in our review below to this chapter that focused on the theme of emotions and science teacher's work. Studies of emotion and science teachers' work are remarkably diverse theoretically and empirically, indicating a thriving topic and posing challenges for the synthesis of findings. Theoretically, studies on science teachers' work are biased towards emotions as individual phenomena. Dominance of individual emotion theory drives methods and findings that attribute emotions to the experiences of individual teachers in response to their internal and environmental contexts.

Empirical outcomes in this chapter point to the role emotions play in teacher and student interactions or relationships, classroom decision-making, development pedagogical content knowledge, shaping classroom climates, teacher wellbeing, career motivation, learning to teach science, effects of prior school experiences on preservice teacher experiences, differences in emotions associated with science subjects (e.g., physics, chemistry, biology), and teaching for social justice. Across these topics, emotions can serve as antecedents to inservice or preservice science teachers' experiences. For example, emotions experienced when science teachers were school students affect dispositions towards teaching science. But emotions were also outcomes of inservice teaching experiences or teacher education classes, such as the joy of observing school students learning science, or enjoying a teacher education course. In addition, emotions were also involved in becoming science teachers by affecting identity or self-efficacy as noted in Chap. 8.

Beyond emotions acting as antecedents and outcomes of science teaching experiences, a number of studies focused on and reported empirical outcomes about the emotional work that science teachers perform. This emotional work can be a source of teacher burnout and stand to undermine longevity in the profession. Studies with this focus identify ways in which the teaching profession and teacher education can best prepare science teachers for reducing the negative impact that emotional work can have.

Overall, few studies focused on the same topics or shared theoretical roots. A clear direction for future research on science teachers' emotions could expand on and replicate the topics and theories already evident in the included studies. However, the field would also benefit from studies adopting theories of emotion that differ from the dominance of individual emotion theories. For example, social theories of emotion that address macro and meso-social forces shaping teachers' work and emotions would be very beneficial for understanding the topics identified in this chapter. Understanding how social structure affords and constrains science teachers' work, shapes emotion, and influences the emotional labor of the job would extend the teachers' work theme beyond the established catalogue of teachers' emotional responses involved with becoming a science teacher or enacting science teaching.

References

NB Articles included in the systematic review are indicated using an asterisk (*)

Aloe, A., Amo, L., & Shanahan, M. (2014). Classroom management self-efficacy and burnout: A multivariate meta-analysis. *Educational Psychology Review, 26*, 101–126.

Bandura, A. (1977). Self-efficacy: Toward a unifying theory of behavioral change. *Psychological Review, 84*(2), 191–221.

*Bellocchi, A., Davis, J., & King, D. T. (2018a). Science demonstrations as mediators of interaction ritual chains: Temporal transitions from mundane to intense emotional energy. In S. M. Ritchie & K. G. Tobin (Eds.), *Eventful learning: Learner emotions* (pp. 57–86). Brill. https://doi.org/10.1163/9789004377912_004

*Bellocchi, A., Mills, K., Olson, R., Patulny, R., & McKenzie, J. (2018b). Emotion work at the frontline of STEM teaching. In L. Bryan & K. Tobin (Eds.), *Critical issues and bold visions for science education: The road ahead* (pp. 247–264). Brill. https://doi.org/10.1163/9789004389663_014

Bhabha, H. K. (1994). *The location of culture*. Routledge.

Billett, S. (2011). *Guidelines for practice: Integrating practice-based experiences*. Strawberry Hills, NSW. http://www.olt.gov.au/resource-integrating-practice-based-experiences-griffith2011.

Borrachero, A. B., Brígido, M., Gómez, R., Bermejo, M. L., & Mellado, V. (2011). The Emotions of the future secondary teachers about teaching and learning of science. *International Journal of Developmental and Educational Psychology. INFAD Revista DePsicología 2*(1), 521–530

*Borrachero, A. B., Brígido, M., Mellado, L., Costillo, E., & Mellado, V. (2014). Emotions in prospective secondary teachers when teaching science content, distinguishing by gender.

Research in Science and Technology Education, 32(2). https://doi.org/10.1080/02635143.2014.909800

*Brick, K., Cooper, J., Mason, L., Faeflen, S., Monmia, J., & Dubinsky, J. M. (2021). Tiered neuroscience and mental health professional development in Liberia improves teacher self-efficacy, self-responsibility, and motivation. *Frontiers in Human Neuroscience, 15*, 664730. https://doi.org/10.3389/fnhum.2021.664730

Brígido, M., Caballero, A., Conde, M. L., Mellado, C., & Bermejo, V. (2009). The emotions in science of trainee student of primary education during teaching practice. *Revistade Educación Campo Abierto, 28*(2), 153–177

*Brígido, M., Bermejo, M. L., Conde, M. C., & Mellado, V. (2010). The emotions in teaching and learning nature sciences and physics/chemistry in pre-service primary teachers. *US-China Educational Review, 7*(12), 25–32.

*Brígido, M., Borrachero, A., Bermejo, M. L., & Mellado, V. (2013a). Prospective primary teachers' self-efficacy and emotions in science teaching. *European Journal of Teacher Education, 36*, 1–18. https://doi.org/10.1080/02619768.2012.686993

*Brígido, M., Couso, D., Gutiérrez, C., & Mellado, V. (2013b). The emotions about teaching and learning science: A study of prospective primary teachers in three Spanish universities. *Journal of Baltic Science Education, 12*(3), 299–311.

*Bullough, R.V. (2009). Seeking eudaimonia: The emotions in learning to teach and to mentor. In Schutz, A., & Zembylas, M. (Eds.), *Advances in teacher emotion research* (pp.33–54). Springer. DI: https://doi.org/10.1007/978-1-4419-0564-2_3

*Dos Santos, F. M. T., & Mortimer, E. F. (2003). How emotions shape the relationship between a chemistry teacher and her high school students. *International Journal of Science Education, 25*(9), 1095–1110. https://doi.org/10.1080/0950069032000052216

*Erdemir, M., & Kandil İngeç, Ş. (2023). Correlation between energy metaphor and empathy tendency. *Journal of Baltic Science Education, 22*(5), 813–832. https://doi.org/10.33225/jbse/23.22.813

Flores, M. A., & Day, C. (2006). Contexts which shape and reshape new teachers' identities: A multiperspective study. *Teaching and Teacher Education, 22*, 219–232.

*González-Weil, C., Reyes-Espejo, M. I., Bravo González, P., Acuña Ruz, F., & Fernández Araneda, F. (2025). How do you trust in the midst of mistrust? A longitudinal study (2010–2021) on building trust in a professional learning community of science teachers in Chile. *Professional Development in Education, 51*(1), 39–53. https://doi.org/10.1080/19415257.2024.2431701

Gross, J. J. (1998). The emerging field of emotion regulation: An integrative review. *Review of General Psychology, 2*, 271–299.

Hargreaves, A. (1998). The emotional politics of teaching and teacher development: With implications for educational leadership. *International Journal of Leadership in Education, 1*(4), 315–336.

Hochschild, A. R. (1983). *The managed heart: Commercialization of human feeling*. University of California Press.

*Hong, M., & Kang, N. H. (2010). South Korean and the US secondary school science teachers' conceptions of creativity and teaching for creativity. *International Journal of Science and Mathematics Education, 8*, 821–843. https://doi.org/10.1007/s10763-009-9188-5

*Kordíková, B., & Brestenská, B. (2022). Bilingual science education: Perceptions of Slovak in-service and pre-service teachers. *International Journal of Bilingual Education and Bilingualism, 25*(3), 1–14. https://doi.org/10.1080/13670050.2020.1718590

Krathwohl, D. R., Bloom, B. S., & Masia, B. B. (1964). *Taxonomy of educational objectives: The classification of educational goals, Hand book II: Affective domain*. David Mckay Company Inc.

*Ma, H. (2018). Influence of the interplay of habitual affective attributes and classroom learning environments on learners' situational affective experiences in learning science: The narratives of primary pre-service teachers. *Research in Science Education, 51*, 399–417. https://doi.org/10.1007/s11165-018-9800-1

*Mahmood, M. S., Talafian, H., Shafer, D., Kuo, E., Lundsgaard, M., & Stelzer, T. (2024). Navigating socio-emotional risk through comfort-building in physics teacher professional development: A case study. *Journal of Research in Science Teaching, 61*, 2223–2257. https://doi.org/10.1002/tea.21949
*Membiela, P., Acosta, K., & González, A. (2023). Antecedents of Chilean in-service science teachers' emotions: The important role of teacher motivation towards teaching career and self-efficacy. *European Journal of Teacher Education, 48*(2), 1–20. https://doi.org/10.1080/02619768.2023.2229003
Moore, F. M. (2008). Agency, identity, and social justice education: Preservice teachers' thoughts on becoming agents of change in urban elementary science classrooms. *Research in Science Education, 38*(5), 589–610.
*Moore Mensah, F. (2019). Finding voice and passion: Critical race theory methodology in science teacher education. *American Educational Research Journal, 65*(4). https://doi.org/10.3102/0002831218818093.
Noddings, N. (1996). Stories and affect in teacher education. *Cambridge Journal of Education, 26*(3), 435–447.
Olson, R., McKenzie, J., Mills, K., Patulny, R., Bellocchi, A., & Caristo, F. (2019). Gendered emotion management and teacher outcomes in secondary school teaching: A review. *Teaching and Teacher Education, 80*, 128–144. https://doi.org/10.1016/j.tate.2019.01.010
Pekrun, R. (2006). The control-value theory of achievement emotions: Assumptions, corollaries, and implications for educational research and practice. *Educational Psychology Review, 18*(4), 315–341.
Pekrun, R., & Perry, R. P. (2014). Control-value theory of achievement emotions. In R. Pekrun & L. Linnenbrink-Garcia (Eds.), *International handbook of emotions in education* (pp. 120–141). Taylor & Francis.
*Praderio Gaias, F. N., Furman, M., Luzuriaga, M., Borrego, E. C., & Ayuso Fernández, G. E. (2024). The influence of emotions on science teaching: A case study with three early childhood education teachers. *Frontiers in Education, 9*, 1472721. https://doi.org/10.3389/feduc.2024.1472721
*Ritchie, S. M., Kidman, G., & Vaughan, T. (2007). Professional learning opportunities from uncovering cover stories of science and science teaching for a scientist-in-transition. *Cultural Studies of Science Education, 2*, 225–242. https://doi.org/10.1007/s11422-006-9044-7
*Ritchie, S. M., Tobin, K. G., Hudson, P., Roth, W.-M., & Mergard, V. (2011). Reproducing successful rituals in bad times: Exploring emotional interactions of a new science teacher. *Science Education, 95*, 745–765. https://doi.org/10.1002/sce.20440
*Rivera Maulucci, M. S. R. (2011). Language experience narratives and the role of autobiographical reasoning in becoming an urban science teacher. *Cultural Studies of Science Education, 6*, 413–434. https://doi.org/10.1007/s11422-011-9323-9
*Rivera Maulucci, M. S. R. (2013). Emotions and positional identity in becoming a social justice science teacher: Nicole's story. *Journal of Research in Science Teaching, 50*, 453–478. https://doi.org/10.1002/tea.21081
Solomon, R. C. (1993). *The passions: Emotions and the meaning of life.* Hacket Publishing Company.
Solomon, R. C. (2004). Emotions, thoughts, and feelings: Emotions as engagements with the world. In R. C. Solomon (Ed.), *Thinking about feeling: Contemporary philosophers on emotion* (pp. 76–88). Oxford University Press.
Solomon, R. C. (2007). *True to our feelings: What our emotions are really telling us.* Oxford University Press.
*Thomson, M. M., & Turner, J. E. (2019). The role of emotions in teachers' professional development: Attending a research experience for teachers (RET) Program. *Education Research International.* https://doi.org/10.1155/2019/5856793
Turner, J. H. (2007). *Human emotions: A sociological theory.* Routledge.

*Uzuntiryaki-Kondakci, E., Kirbulut, Z. D., Oktay, O., & Sarici, E. (2022). A qualitative examination of science teachers' emotions, emotion regulation goals and strategies. *Research in Science Education, 52*, 1131–1155. https://doi.org/10.1007/s11165-020-09986-y

van Aalderen-Smeets, S. I., Walma van der Molen, J. H., & Asma, L. J. F. (2012). Primary teachers' attitudes toward science: A new theoretical framework. *Science Education, 96*(1), 158–182. https://doi.org/10.1002/sce.20467

Weldon, P. (2018). Early career teacher attrition in Australia: Evidence, definition, classification and measurement. *Australian Journal of Education, 62*(1), 61–78.

*Wicaksono, A. G. C., & Korom, E. (2023). Attitudes towards science in higher education: Validation of questionnaire among science teacher candidates and engineering students in Indonesia. *Heliyon, 9*(9), Article e20023. https://doi.org/10.1016/j.heliyon.2023.e20023

*Zembylas, M. (2001). A paralogical affirmation of emotion's discourse in science teaching. In A. Barton & M. Osborne (Eds.), *Teaching science in diverse settings: Marginalized discourses and classroom practice* (pp. 99–128). Peter Lang.

*Zembylas, M. (2002). Constructing genealogies of teachers' emotions in science teaching. *Journal of Research in Science Teaching, 39*, 79–103. https://doi.org/10.1002/tea.10010

*Zembylas, M. (2004a). Emotion metaphors and emotional labor in science teaching. *Science Education, 88*, 301–324. https://doi.org/10.1002/sce.10116

*Zembylas, M. (2004b). Emotional issues in teaching science: A case study of a teacher's views. *Research in Science Education, 34*, 343–364. https://doi.org/10.1007/s11165-004-0287-6

*Zembylas, M., & Buhlman Barker, H. (2002). Preservice teacher attitudes and emotions: Individual spaces, community conversations and transformations. *Research in Science Education, 32*, 329–351. https://doi.org/10.1023/A:1020862000107

Chapter 10
Conclusions and Future Directions for Research on Emotion in Science Teacher Education

In this systematic review of preservice and inservice science teachers' emotions, we aimed to understand the research landscape by examining topics of research, empirical findings, as well as the theories of emotion and methodologies adopted by researchers. The following research questions guided us:

1. What topics associated with emotions have researchers addressed in the context of preservice and inservice science education?
2. What are the main outcomes from research on emotions in preservice and inservice science education?
3. How are emotions theorized in research on emotion in preservice and inservice science teaching?
4. What methods have been used in emotions research in preservice and inservice science education?

Chapter 3 provided a broad overview of publication patterns addressing Research Questions 1, 3, and 4. Overall, we found a growth in publications from the earliest recorded in our data set appearing in 1995, with the most recent articles in 2025. This growth was not linear, with very clear peaks in volume of outputs in 2013 and 2019, each followed by steep declines; however, an upward trend is clear, yielding an average publication rate of 3.2 articles per annum. More importantly, 26 countries were represented in the study samples, with distribution across Western and Eastern contexts. Despite this geographic distribution, a limitation of the samples is that they are mostly concentrated in a few countries, and dominated by three countries in Western contexts, which limits knowledge in the field across different cultural and science education settings.

In response to Research Question 1, Chaps. 4–9 presented thematic topics evident in our data set of 92 publications: Classroom emotional climate (n = 4; 4.3%); Science inquiry, laboratory, and practical work (n = 17; 18.5%); Controversial issues (n = 15; 16.3%); Technology (n = 8; 8.7%); Science teachers' self-efficacy and identity (n = 19; 20.7%); and Science teachers' work (n = 29; 31.5%). These six thematic

A. Bellocchi et al., *Emotion Research in Science Teacher Education*, Sociocultural Explorations of Science Education 32, https://doi.org/10.1007/978-3-032-07711-0_10

topics indicate that research on preservice and inservice science teachers' emotions is diverse, however, many topics that form the broader interest of science education (e.g., conceptual change, argumentation, activism, representations) have not been addressed, which presents opportunities to consolidate the topics in our themes that have low numbers of articles, as well as opportunities to expand future research into new topics. The distribution of studies (shown in parentheses) within each topic reflects imbalances, and the overall number for the better-represented topics is still minor compared to established traditions of science education research (see also Chap. 3; Lin, 2025; Odden et al., 2021; Tosun, 2024). Limited connectivity between studies within topics due to theoretical and methodological differences complicates the field as we will discuss later when addressing Research Question 2.

Given the increasing importance of technology in learning to teach science and science instruction, more studies are needed. The rise and ubiquity of machine learning in the various Generative Artificial Intelligence software available at the time of writing suggests that this is a fruitful area requiring immediate attention. Across the topics of inquiry represented in our themes, the main focus was on specific emotions (i.e., happiness, frustration, anger, etc.), with fewer giving consideration to collective forms of emotion such as emotional climate. Due to the importance of social context in becoming a science teacher and engaging in science teaching, it is important to conduct further work on these collective emotional experiences. Future studies addressing the above limitations will contribute significantly to our understanding of emotions in science teacher education.

We have summarized the empirical outcomes of included studies in detail in Chaps. 4–9, answering Research Question 2, and it is impossible to include a more general discussion here due to the diversity and volume of outcomes. At a broad level, the outcomes reveal that emotions are implicated in diverse aspects of teaching science and learning to teach science. Overall, empirical findings reveal a nuanced picture of the role that pleasant *and* unpleasant emotions play in teaching science and learning to teach science. Emotions can be antecedents and outcomes of science teachers' experiences, which are shaped by a range of social/relational factors including mentoring, peer support, and teacher-student relationships. When teachers experience emotional barriers or unpleasant emotions associated with teaching science, professional development and careful planning are ways of overcoming these hurdles. Individual differences such as personal backgrounds and preferences for specific science disciplines (i.e., physics, biology, chemistry etc.) are associated with different emotional outcomes for science teachers. Reviewed studies also show that emotions shape science teacher decision-making such as pedagogical choices and classroom practices.

Pleasant and unpleasant emotions are integral to the development of science teacher self-efficacy and identity, with variable effects resulting from experiences of pleasant and unpleasant emotions during science teachers' formative educational experiences in their own schooling. This points to the need for science education researchers and science teachers to develop deep understandings of the role that emotions play in their formation as science teachers. The emotions experienced by science teachers, like all teachers, in relation to the nature of their work has important effects on longevity in the profession. Emotional exhaustion and burnout seem to be perennial problems that could benefit from intervention studies. However, pleasant emotions associated with students and collegial support are supportive in the career. A clear omission in the field is systematic research addressing the emotions of out-of-field science teachers as a unique cohort.

Science inquiry, laboratory and practical work are the source of pleasant and unpleasant emotions for preservice and inservice science teachers alike, and emotions from the teachers' past school science experiences are formative in identity development related to science teaching. It is perhaps unsurprising to find that controversial issues inspire science teachers' emotions, especially when they conflict with strongly held personal views. Given the important personal emotional conflict teachers can experience when controversial topics are the basis for instruction (e.g., SSI) and curriculum content (i.e., evolution), or when such issues arise from student perspectives, controversial issues is an important topic that would benefit from more emotions research. The topic represents only 16% of included studies, and yet, poses some of the more emotionally challenging terrain for science teachers to traverse. Across all research topics identified through our, there is scope for developing a greater number of practical applications from the knowledge developed by researchers in this field. In this way, the field has a research-based foundation for practical implementation; however, the small number of studies per topic, the small number of topics represented in the included studies (compared to the broader range of topics that form the focus of science education research), and exclusion of emotion from recent and past reviews of the field, indicates a need to expand this important and valuable, as this volume has shown, field of inquiry (cf. Lin, 2025; Odden et al., 2021; Tosun, 2024).

Absent from the included studies are those that consider emotion across macro-, meso-, and micro-social levels. At the macro level, we would benefit from studies that consider emotion in relation to science education as social institutions. One approach might be the use of policy analysis to identify tacit and explicit presence or absence of emotion. Macro-social theories of emotion, such as those developed by Kemper (2006), which deal with social structure and power (see rare examples such Moore Mensah 2008 and Zembylas, 2001) are rarely adopted yet would prove beneficial to the field by connecting individual emotion and social structure. At the meso-level, units such as schools might be the focus, and again, analysis of school policy, code of conduct, and other governing documents could be fruitful to understand the structural dimensions of being a science teacher that shape and are shaped by emotion. To offer one example, social discourse in the media, research, and policy surrounding crises in the STEM pipeline could be examined as forms of anxiety. Research could interpret the effects of anxiety-fueled public and research discourses on science education policy and practice.

In response to Research Question 3, the thematic reporting of our review of studies in Chaps. 4–9 revealed that research on emotions in science teacher education ranges from studies that adopted specific emotion theories to those that we classify as atheoretical. Psychological theories dominate the field (37% of all studies), with considerable diversity in the specific theories applied as reported in Chap. 3 and Table 4. As discussed in our brief overview of emotion theory in Chap. 1, when psychology turned its attention to emotions as a serious topic of inquiry around the 18th Century, scholarship grew and the topic was cemented in the field, as well as in public imagination, undoubtedly through the evolution of clinical practices and popular psychology. In addition to the dominance of psychological theories in our data set, within these perspectives, individual psychology was more prevalent than social psychological theories. This presents a limitation as social psychological theories of emotion have much to offer in the interpretation of science teachers' experiences given the high degree of socialization in the job. Sociology is the second largest *disciplinary* source of theory, though its proportion (9%) is meagre compared to psychological theories. This exacerbates the limitation noted before regarding the prevalence of individual psychological theories of emotion as numerous theories from the sociology of emotion have social life at the center and offer much for the interpretation of science teachers' work and learning to become a teacher. More broadly in emotions research, cultural studies perspectives (e.g., Ahmed, 2010) have gained popularity, but these are yet to find strong purchase in science teacher education research. However, interested scholars might heed Lively's (2024) observation that there may not be substantial differences in such approaches and the sociology of emotion. Overall, increasing the number of studies adopting theories with a social rather than individual focus would be beneficial.

In preparing this volume, we were humbled by the value that the history and philosophy of emotion has as a metatheoretical lens for interpreting emotions across the literature (c. Dixon, 2012). Although we touched briefly on history and philosophy of emotion in Chap. 1, certain ideas, such as emotion/reason and emotion/cognition dichotomies, have been repeated across time and scholarship. It was interesting to find in our review that sime scholars challenged emotion/reason dichotomies, whereas others reproduced these dichotomies. Engagement with contemporary emotion theory or the history and philosophy of emotion can potentially avoid rehashing bygone debates if the field has reached a consensus. This point also stands in support of our more general argument that intentional adoption of emotion theory is beneficial for a field that includes intentional atheoretical approaches (elaborated later).

Moving forward, researchers could consider the relevance of theoretical choice for the cultural context of the chosen samples. More studies adopting theories from Eastern foundations to study samples from Eastern contexts would enrich the field, and clear and intentional alignment between theory and context is generally needed. One included study acknowledged the potential limitations of adopting Western philosophical ideas about emotions when interpreting Eastern contexts, an important next step for future inquiry (cf. Rinchen et al., 2016) and a more recent study adopted cultural labels for emotions relevant to the participant sample (i.e., Koul et al., 2024). A greater number of studies in countries with different philosophical and cultural foundations would improve the field by providing better understandings of the role different cultures play in defining and shaping science teachers' emotions.

Moreover, research that compares the value of Eastern and Western theoretical perspectives on emotions would serve to advance the field. Although Scarantino (2024), among other scholars, has identified common elements across cultures about emotion concepts, there is also merit to the constructionist argument about cultural specificity of emotion labels and concepts (cf. Feldman-Barrett, 2017). That is, English-language emotion labels appeared across studies in our review from a diverse range of cultural contexts. At times, the labels were imported methodologically by adopting instruments designed with English emotion labels, that were translated into the language of the sample country. However, Feldman-Barrett (2017) notes that even within one macro-cultural group (i.e., Eastern/Western) there are emotion labels used in some countries that have no equivalent in other languages. An argument we *cannot* sustain from our review is that there is a need for more theories to be *developed* or for more author-generated conceptual frameworks to be propagated. It appears that emotion research writ large would benefit from more studies across fields to validate or refute existing theories or their concepts, rather than proliferating them.

A considerable number of studies did not include emotion theory with a frequency of 32%, making this the second largest category in our reviewed studies. We identify three atheoretical approaches that we name as follows: 1) lay emotions; 2) literature-informed emotions; and 3) methodological emotions. *Lay emotions* refers to studies that use the term *emotion,* perhaps based on an everyday understanding of the term, without reference to a theory or disciplinary scholarship and definitions of emotion. Such studies do not state any intentional use of the term, leading us to infer from terminology used within the works that lay perspectives are operative. Studies adopting lay perspectives tend to be those where emotion is an incidental focus of the research rather than emotion forming the central phenomenon of investigation. In addition, there are a small number of studies that presented *literature-informed* approaches to emotions, without stipulating preference for one theory to define or interpret emotions from the data. Such articles do not report if the authors were seeking to develop a conceptual framework on emotions by drawing concepts and constructs from different theories that are then to be used analytically, or if they were simply presenting a review of theories or concepts with no intention of applying these concepts and constructs for data analysis and interpretation. Like the second approach, studies adopting the third approach—*methodological emotions*—present a review of theories, however, one theory, among several that are discussed in the literature review, dominates the methodological approach to the study. For example, this is evident when one theory informs a quantitative survey design where questionnaires contain items aligned with the theory, or the results of questionnaires are analyzed through emotion concepts and constructs from that theory (e.g., see Alarcón-Orozco, et al., 2025; Chap. 4).

Although there are justifiable reasons for researchers to adopt an atheoretical stance, most works taking this approach did not explain whether it was an intentional research design choice, or whether emotions arose as unintentional phenomena in studies focused on other topics. Most of studies, included in this review, that took an atheoretical approach were not aligned with research designs that supported the approach. Without justification for these methodological and theoretical choices, it becomes difficult to integrate findings or evaluate research merit. As Bellocchi and Amat (2022) note, there are social research approaches that align with atheoretical methods, but these research design choices need to be clearly reported.

The absence of any theoretical perspective presents two important challenges. The first is that everyday perspectives can perpetuate antiquated understandings of emotion that do not reflect current emotion scholarship (Bellocchi & Amat, 2022). Moreover, the lay perspective, on the surface, may give the appearance that a normative concept of emotion is the focus of investigation and that researchers are interested in understanding how science teachers might conceptualize emotion in relation to their experiences (cf. Scarantino, 2024). However, studies in our review adopting the lay approach rarely expressed the intent to understand emotion in-and-of itself. Rather, emotion was an object of inquiry more aligned to a natural kind concept, which is best captured through formal theory. The second limitation to the lay approach is that it is best suited to a study design that is concordant with an atheoretical stance towards their central phenomena, such as phenomenology, ethnomethodology, and poststructuralist perspectives (Bellocchi & Amat, 2022). Notwithstanding this critique, it is possible that some studies classified under the atheoretical category were adopting a generic qualitative inductive approach to inquiry. However, without authors explicating such an approach, as reviewers we were left to guess what aspects of study design were intentional and those that were unintentional.

In response to Research Question 4, this review found that, methodologically, research on science teachers' emotions is predominantly qualitative ($n = 47$), with fewer quantitative ($n = 25$), mixed-methods ($n = 4$), and qualitative + quantitative ($n = 16$) studies. Like theoretical choices, methodological choices constrain the phenomena that can be accessed, indicating a need to balance the methods used in understanding emotions. Although mixed-methods research has gained popularity, it was sometimes unclear whether studies adopted an intentional mixed-methods design (e.g., convergent, explanatory, exploratory) or used multiple methods that included quantitative and qualitative data generation and analysis. Another methodological weakness evident in some studies, which can be easily improved, is the clear specification of participant recruitment and sampling methods, particularly for qualitative studies, but also in the case of quantitative studies where the intended population was not clear. Related to sampling, clearer statements about the intended forms generalization (e.g., external generalization, internal generalization, theoretical generalization) would clarify how authors see their work being meaningful beyond the sample and context of the study. Detailed description of context and achieved samples, where appropriate, would also enhance transferability. As most studies focused on either preservice or inservice science teacher education, longitudinal designs that follow emotional trajectories from preservice to inservice teaching would be valuable. We acknowledge that this is a challenging design to implement, but were possible, it would enhance our understanding of emotions across this important time of transition in becoming science teachers.

The field would benefit from methodological articles explaining the nature of the relationship between the methods chosen (e.g., interviews, observations, questionnaires) and the emotion theories adopted. In the reviewed research, it was sometimes unclear how specified theories of emotion (e.g., neuroscience) were relevant for data analysis involving methods (e.g., interviews) that did not access emotions as suggested by the theory. Returning to Scarantino's (2024) *natural kind* and *normative kind* concepts, one might expect researchers in the natural kind tradition to adopt theories about emotion that, for example, center on physiology, and methods that capture physical changes in the body. Interviews using questions that uncover the meaning a person gives for an emotional experience would align with a normative approach. A theory that can manage the discursive construction of emotion concepts could be helpful for analyzing such data. In addition, careful alignment between theory, study design, analyses, and claims would strengthen science teacher emotion research in general.

General trends related to study characteristics (i.e., Table 3; Chap. 3) present other limitations of emotion research in science teacher education. Most studies focused on preservice teachers in elementary contexts within qualitative research designs. US teachers were the most represented cohort, with Turkey and Australia second and third, each with about half the number of studies as the US. This distribution might differ if articles in other languages were included. For instance, Spain had the fourth highest number of studies. Scholars from Spain frequently referred to articles published in Spanish-language journals, which were omitted by our inclusion criteria. Despite these limitations, 26 countries were represented in our data set, demonstrating clear international interest in this field, albeit often represented by isolated studies. Aside from the cohort limitation this presents for purposes of external generalization about emotion outcomes across the globe, any practical recommendations currently in circulation, are likely to be based on understandings of emotion and science education from a narrow range of countries. Cultural sensitivity and local education needs differ vastly, and existing research may have limited transferability internationally. One key feature of research that supports transferability is detailed information about the context in which research takes place. Although at times researchers may wish to avoid providing contextual details to protect the identity of participants (especially in studies with small samples), we did encounter some studies where better description of the context would have supported transferability of the findings.

Our review is not without limitations. Most important of which is the inclusion criteria used for selecting articles. We noted above the impact that choosing only English language articles for our data set presents cultural limitations for understandings of emotion. The choice also reduces true indications of the volume of articles on science teachers' emotions, the range of cohorts studied, topics of inquiry (our themes), and overall empirical knowledge about emotions in teaching science and learning to become a science teacher.

Another key limitation, although this time resulting from the studies in the field, is the challenge of synthesizing empirical outcomes when it is not clear if different researchers address emotions in the same way. The main contributors to this limitation are the theoretical matters we have discussed above; however, lack of details about samples and sampling, research design, and methods used render the task more challenging. Lastly, we did not apply research quality criteria in the selection or exclusion of articles. Due to the relatively low number of publications on emotion in the field, and a desire to map knowledge about emotion in science teacher education, we reviewed studies that may have been included should quality criteria, especially centered on research design, have been applied. Our approach to reviewing each included article in-depth, reduces potential adverse effects of this limitation as readers can make up their own minds about the veracity of empirical outcomes produced by studies where doubt remains around theoretical and methodological choices.

As the focus of our review was to understand elements of emotion research (e.g., theory, methods) and empirical outcomes, this volume does not do justice to the range of practical implications offered by authors of included studies. Future reviews could synthesize practical implications derived from the data set in this review to present a complementary volume that guides practice.

Our goal in this concluding chapter was to summarizes the key trends in research on emotion in science teacher education identified through the systematic review of literature to identify broad limitations to research in the field, and suggest future directions. Research directions specific to each thematic topics were addressed in the respective chapter and were not repeated here (i.e., Chaps. 4–9). In presenting a critique of limitations in current research, we have not lost sight of the fact that most reviewed studies present high-quality research that, across the collection of included studies, spans diverse social research epistemologies, theories, and methods. These exemplary works will be self-evident to readers in the thematic chapters, which, with the affordances of the longer written format, allow us to do justice to details of the studies.

We have reported that emotions research in science teacher education is a lively and diverse field with widespread and growing international interest. We have provided directions for future research to advance our global understanding of this increasingly important topic. As the histories of emotion, and percussor concepts show, emotions are the essence of our humanity and have held scholarly interest since antiquity. In a field of inquiry such as science education, which works with, about, and for teachers and students, humane practices and understandings cannot exist without attention to such an essential aspect of our being. We hope this volume encourages others to join the growing tide.

References

NB Articles included in the systematic review are indicated using an asterisk (*)

Ahmed, S. (2010). *The promise of happiness*. Duke University Press.

*Alarcón-Orozco, M. M., Franco-Mariscal, A. J., Oliva, J. M., & Blanco-López, Á. (2025). Emotions experienced by preservice early childhood teachers during a training program in inquiry-based science education. *Journal of Research in Science Teaching*. https://doi.org/10.1002/tea.70007

Bellocchi, A., & Amat, A. (2022). Emotion and science teacher education. In J. A. Luft, & M. G. Jones, (Eds.), *Handbook of research on science teacher education* (pp. 426–438). Routledge. https://doi.org/10.4324/9781003098478-38

Dixon, T. (2012). *From passions to emotions: The creation of a secular psychological category*. Cambridge University Press.

Feldman-Barrett, L. (2017). *How emotions are made: The secret life of the brain*. Macmillan.

Kemper, T. D. (2006). Power and status and the power-status theory of emotions. In J..E. Stets & J. H. Turner (Eds.), *Handbook of the sociology of emotions*. Springer. https://doi.org/10.1007/978-0-387-30715-2_5

*Koul, R., Musikul, K., Nishikawa, S., & Almutawa, H. (2024). Mentoring in the development of science teaching self-efficacy among primary school teachers in Thailand: A Mixed Methods study. *Journal of Science Teacher Education, 35*(1), 44–62. https://doi.org/10.1080/1046560X.2023.2179384

Lin, T. J., Lin, T. C., Potvin, P., Tsai, C-C. (2025) Research trends in science education from 2018 to 2022: A systematic content analysis of publications in selected journals. *International Journal of Science Education, 47*(4), 510–538. https://doi.org/10.1080/09500693.2024.2340811

Lively, K. J. (2024). An overview of contemporary theories of emotions in sociology. In A. Scarantino (Ed.), *Emotion theory: The Routledge comprehensive guide: Volume 1: History, contemporary theories, and key elements* (pp. 287–309). Routledge. https://doi.org/10.4324/9781315559940-17

Moore, F. M. (2008). Agency, identity, and social justice education: Preservice teachers' thoughts on becoming agents of change in urban elementary science classrooms. *Research in Science Education, 38*(5), 589–610.

Odden, T. O., B., Marin. A., & Rudolph, J. L. (2021). How has *Science Education* changed over the last 100 years? An analysis using natural language processing. *Science Education, 105*, 653–680. https://doi.org/10.1002/sce.21623

*Rinchen, S., Ritchie, S. M., & Bellocchi, A. (2016). Emotional climate of a pre-service science teacher education class in Bhutan. *Cultural Studies of Science Education, 11*, 1–26. https://doi.org/10.1007/s11422-014-9658-0

Scarantino, A. (2024). Introduction to volume 1: The value of history, a wealth of theoretical options, and the elements of emotion theory. In A. Scarantino (Ed.), *Emotion theory: The Routledge comprehensive guide: Volume 1: History, contemporary theories, and key elements* (pp. 1–13). Routledge. https://doi.org/10.4324/9781315559940-1

Tosun, C. (2024). Analysis of the last 40 years of science education research via bibliometric methods. *Science & Education, 33*, 451–480. https://doi.org/10.1007/s11191-022-00400-9

*Zembylas, M. (2001). A paralogical affirmation of emotion's discourse in science teaching. In A. Barton & M. Osborne (Eds.), *Teaching science in diverse settings: Marginalized discourses and classroom practice* (pp. 99–128). Peter Lang.

MIX
Papier aus verantwortungsvollen Quellen
Paper from responsible sources
FSC® C105338

If you have any concerns about our products,
you can contact us on
ProductSafety@springernature.com

In case Publisher is established outside the EU,
the EU authorized representative is:
Springer Nature Customer Service Center GmbH
Europaplatz 3, 69115 Heidelberg, Germany

Printed by Libri Plureos GmbH
in Hamburg, Germany